Heart Failure in Animals
Current Concepts

NIPA GENX ELECTRONIC RESOURCES & SOLUTIONS P. LTD.
New Delhi-110 034

Prof. (Dr.) J.P.Varshney, Former Principal Scientist (Veterinary Medicine, ICAR-IVRI Izatnagar), is presently working as Senior Consultant (Veterinary Medicine) at Nandini Veterinary Hospital, Surat (Gujarat). He is a clinical scientist of repute having a vast professional experience in the area of cardiology, canine internal medicine, exotic pet medicine, dietary disorders of ruminants, homeopathy in veterinary medicine, mastitis, and equine medicine. He had been a visiting faculty in the area of cardiology in veterinary colleges. He has been bestowed with Dr. K.S. Nair Memorial Gold Medal-1990, Dr. C.G. Bhaskar Gold Medal—1996, Ram Lal Agrawal Gold Medal-1997, Award of Merit-2001, Award of Honour-2002, Best Teacher Award 2003-2004 (IVRI, Izatnagar), PETCARE Award for Canine Excellence-2005 and Dr. C.M. Singh Memorial Lifetime Achievement Award-2020 besides many other awards. Apart from publications in national/ international scientific journals of repute, chapters in books, bulletins and monographs, his books on “Equine Parasitic Diseases and Their Management (ICAR-IVRI Izatnagar Publication, 2003),“Research Findings -Homeopathic Bio- efficacy and Management of Animal Health (Sintex International limited, Kalol, 2007) ,“Electrocardiography in Veterinary Medicine” (Springer Nature, Singapore, 2020), “Handbook of Exotic Pet Medicine” (Brillion Publishing, New Delhi, 2021), and “Ultrasound in Veterinary Medicine: Fundamentals and Applications “(NIPA, New Delhi, 2022) are the testimony of his clinical acumen and contributions in the field of clinical veterinary medicine.

Heart Failure in Animals Current Concepts

J.P. Varshney

B.V.Sc. and A.H., M.V.Sc. (Medicine); Ph.D. (Vet. Med.)
Former Principal Scientist
Department of Veterinary Medicine
ICAR-Indian Veterinary Research Institute, Izatnagar, Uttar Pradesh
and
Senior Consultant (Veterinary Medicine) Nandini Veterinary Hospital
Ghod-Dod Road, Surat-395001, Gujarat

NIPA GENX ELECTRONIC RESOURCES & SOLUTIONS P. LTD.
New Delhi-110 034

NIPA GENX ELECTRONIC
RESOURCES & SOLUTIONS P. LTD.

101,103, Vikas Surya Plaza, CU Block
L.S.C.Market, Pitam Pura, New Delhi-110 034
Ph : +91 11 27341616, 27341717, 27341718
E-mail:newindiapublishingagency@gmail.com
www: www.nipabooks.com

For customer assistance, please contact
Phone: + 91-11-27 34 17 17 Fax: + 91-11- 27 34 16 16
E-Mail: feedbacks@nipabooks.com

ISBN: 978-93-94490-34-5

Composed and Designed by NIPA.

Dedicated
To
My Beloved Late Parents
And
Teachers
For
Shaping My Life

Acknowledgements

Words are not enough to express my hearty feelings to my life partner Mrs. Jai PrabhaVarshney for her sacrifices, unconditional support, positive thinking, extraordinary care, encouragement and patience. I take this opportunity to thanks my other family members Dr. Jai PratibhaVarshney (daughter) and Brig. (Dr.) Atul Seth (son-in-law); Jai Prabhat Varshney (son) and Mrs. Ritu Gupta (daughter-in –law); Mrs. Jai Prabhanshi Varshney (daughter) and Ajit Varshney (son-in-law); and grand children (Agrima Seth, Vedaant Varshney, DevanshVarshney and Anvi Varshney) who always care and support me.

Thanks are also due to Shri Nayan N. Bharatia Managing Trustee and Board of Trustees, Nandini Veterinary Hospital, Surat for encouragement and support. Direct or indirect help rendered by the collogues and staff of the Hospital is duly acknowledged.

(J.P. Varshney)

Preface

The word "Heart Failure" is scaring to a vast majority of peoples. It is the terminal event of heart diseases and is a complex progressive clinical syndrome of symptoms and signs caused by impairment of heart pumping action owing to structural and /or functional abnormalities. Though all species of animals have heart problems and some of these may lead to heart failure, it is of more concern in companion animals. India has around 20 million pet dogs and 2 million pet cats besides a vast number of stray animals. Magnitude of the problem is reflected by estimates that approximately 10.0 percent dogs suffer from some form of heart disease. With the changed life style, heart diseases are being recognized in pet companions with greater frequency. All heart diseases may not end up in heart failure. Nevertheless, prevalence of heart failure in canine and feline is escalating. Heart failure is becoming a worldwide problem not only in companion animals but also in humans. Because of increasing prevalence of heart diseases in companion animals, cardiology is emerging as a specialty in veterinary medicine.

A majority of the practicing veterinarians are not adequately equipped with diagnostic and therapeutic skills to deal with heart failure in different species of animals. Lack of understanding of pathogenesis of heart failure, meagre information about application and interpretation of diagnostic modalities, and scanty knowledge about cardiac drugs and their interactions impede desirable outcomes.

The demonstration of objective evidence of cardiac structural or functional abnormalities is necessary for the diagnosis of heart failure to be made. Symptoms most commonly encountered in dogs with heart failure are breathlessness/ panting/dyspnea/orthopnea/nocturnal coughing and exertion. These symptoms are not exclusive to heart failure but can also be exhibited in pulmonary diseases creating a confusing state. An early detection of heart failure is very crucial for timely pharmacological interventions to slow down the disease progression and to improve quality of life and treatment outcomes. But an early detection of cardiovascular disease is a challenging task in traditional approach. No single diagnostic test is completely perfect for diagnosing heart failure and therefore diagnosis relies on systematic clinical assessment based on a combination of history, clinical examination and cardiac appropriate investigations. Recently some novel techniques such as echocardiography, cardiovascular magnetic resonance (MR), nuclear imaging- positron emission tomography(PET),single-photon emission computed tomography (SPECT), enzymes, hormones, biological substances, heart rate variability and so on have become available for an early detection of heart failure in humans. Some of these techniques are also available now in animals. But their application and interpretation is challenging due to differences in cardiac size and rate requiring validation and proper training.

Our understanding of heart failure in canine and feline has evolved tremendously over last few decades. Recent studies have indicated significant increase in C-reactive protein (CRP), total leukocytes and neutrophil counts in dogs with severe decompensated heart failure indicating systemic inflammation. Hence increase in the values of C-reactive protein, leukocytes and neutrophils in dogs with heart failure should not be construed as infection. Evidence is also accumulating about an increase in pro-inflammatory cytokines and oxidative stress in many cases of heart failure.

Keeping pace with developments in heart failure will facilitate better management to increase longevity and quality of life of animals with heart failure. In the past, treatment of heart failure in dogs relied heavily on digoxin, furosemide, and ACE-inhibitors like enalapril. Now we can offer more advanced therapies that can assists dogs to live longer with a better quality of life. These days, heart failure is treated with drugs meant to reduce blood pressure, improve myocardial contractility, control arrhythmias , reduce remodeling of myocardial muscles, do not increase myocardial demands for oxygen, and do not increase calcium in the myocardial cells. During last two decades drugs like pimobendan, benezapril, spironolactone have been added in the therapeutic protocol of heart failure in dogs. New drugs for new targets in heart failure with minimum side effects are being continuously researched for increasing longevity and better quality of life of heart patients. Gene therapy, stem cell therapy, chronopharmacotherapy, identification of new targets and development of drugs to block these new targets, and surgical interventions may revolutionize heart failure treatment in coming years.

Present scenario in India reflects dearth of veterinary cardiologist to cope up the increasing demand of handling cardiac patients in animals specially companion animals (dogs, cats, pet avian etc.). The existing multitasked veterinarians are not well conversant with intricacies of heart failure and its diagnostic and therapeutic approaches. Books dealing with all these aspects are also not easily accessible to veterinarians in India.

The present book “Heart Failure in Animals: Current Concepts” is an endeavor to provide comprehensive information about structure and functions of the heart, current concepts in etiopathogenesis of heart failure, updated diagnostic modalities appropriate to heart failure, therapeutic interventions, new pharmacological agents and dietary considerations in heart failure of dogs. Description of related syndromes like syncope, shock and cardiac arrest has widened the horizon. Chapters on heart failure in feline, ruminant, equine, leporine ,pet avian, ferret and man have given a comparative outlook of heart failure in these species and have broadened the utility and scope of the book to veterinary students, researchers, specialist in veterinary medicine as well as practitioners in equipping them to handle heart failure cases in a more effective and rational way.

Surat, Gujarat, India **J.P. Varshney**

Contents

List of Abbreviations

AC-6	Adenylate Cyclase 6
ACE-I	Angiotensin Converting Enzyme- Inhibitor
ACTH	Adrenocorticotropic hormone
@	At the rate of
ALS	Advance Life Support
APC	Atrial Premature complex
ARNI	Angiotensin Receptor Neprilisyn Inhibitor
ASIs	Aldosterone Synthase Inhibitors
AST (SGOT)	Aspartate aminotransferase (serum glutamic- oxaloacetic transaminase)
ATA	Atmospheres Absolute
AT-I	Angiotensin-I
AT-II	Angiotensin-II
ATP	Adenosine Triphosphate
AV block	Atrio-ventricular Block
AV node	Atrio-ventricular node
AV valves	Atrio-ventricular valves
BCHF	Bovine Congestive Heart Failure
BID	Bis in die (two times a day)
BLS	Basic Life Support
BNP	Brain Natriuretic Peptide
BP	Blood Pressure
BPM	Beat Per Minute
BUN	Blood Urea Nitrogen
BZD	Benzodiazepine
^{o}C	Degree Centrigrade
Ca++	Calcium
CA	Cardiac Arrest
CBC	Complete Blood Count
CHF	Congestive Heart failure
CPK	Creatine Phosphokinase
cm	Centimeter
CMR	Cardiovascular Magnetic Resonance
CMVI	Chronic Mitral Valve Insufficiency
CNS	Central Nervous System
COP	Capillary Oncotic Pressure
CPA	Cardio-pulmonary Arrest
CPCR	Cardio-pulmonary Cerebral Resuscitation
CPR	Cardio-pulmonary Resuscitation
CRI	Constant Rate Infusion
CRP	C-Reactive Protein

CRT	Cardiac Resynchronization Therapy
CT	Computed Tomography
cTn-I	Cardiac Troponin-I
CUPRAC	Cupric Reducing Anti-oxidant Capacity
CVP	Central venous Pressure
DCM	Dilated cardiomyopathy
DHA	Docosahexaenoic Acid
D. immitis	Dirofilaria immitis
DM	Dry Matter
DMVD	Degenerative Mitral valve Disease
DNA	Deoxyribonucleic acid
DV	Dorso- Ventral
E.coli	Escherichia coli
E.cuniculi	Encephalitozoon cuniculi
ELISA	Enzyme Linked Immunoassay
ECG/EKG	Electrocardiogram
EPA	Eicosapentaenoic Acid
EtCO2	End Tidal Carbon Dioxide
0 F	Degree Fahrenheit
FDA (USA)	Food and Drug Administration (United States of America)
Fi O2	Fraction of inspired Oxygen
FMD	Foot and Mouth Disease
g	Gram
GABA	Gamma Amino-Butyric Acid
GAGs	Glucosaminoglycans
GI	Gastro-intestinal
GM-CSF	Granulocyte Macrophase Colony Stimulating Factor
GRs	Glucocorticoid Receptors
H2O	Water
Hb	Haemoglobin
HBOT	Hyperbaric Oxygen Therapy
HCM	Hypertrophic Cardiomyopathy
HF	Heart Failure
HF-rEF	Heart Failure with Reduced Left Ventricular Ejection Fraction
HF-pEF	Heart Failure with Preserved Let Ventricular Ejection Fraction
Hg	Mercury
hPSCs	Human Pluripotent Stem Cells
hr	Hour
5-HT	5-Hydroxytryptamine
5-HT2	5-Hydroxytryptamine 2
5-HT3	5-Hydroxytryptamine 3
ICDs	Implantable Cardioverter Defibrillators
i.e.	That is
IFN	Interferon
IL	Interleukin
IM	Intramuscular
IO	Intraosseous
IPPV	Intermittent Positive Pressure Ventilation
IT	Intratracheal

IV	Intravenous
K+	Potassium
KC like	Keratinocyte Chemotactic- Like
Kg	Kilogram
<	Less than
l/L	Liter
L	Left
LA	Left Atrium
LDH	Lactic Dehydrogenase
LE	Lupus Erythromatous
LH	Left Heart
LV	Left Ventricle
LVOTO	Left Ventricular Outflow Tract Obstruction
>	More than
MCP-1	Monocyte Chemotactic Protein-1
mEq/l	Mille equivalent /liter
MI	Myocardial Infarction
MIMI	Microscopic Intramural Myocardial Infarction
Min.	Minute
mg	Milligram.
mm	Millimeter
MMVD	Myxomatous Mitral Valve Disease
MPS VII	Muco Polysaccharidosis VII
MR	Mitral Regurgitation
MR	Magnetic Resonance
MRI	Magnetic Resonance Imaging
MRs	Mineralo corticoid Receptors
mV	Milli Volt
Na+	Sodium
NADPH	Nicotinamide Adenine Dinucleotide Phosphate
NSAIDs	Non-steroidal Anti-inflammatory Drugs
NT-pro BNP	N-Terminal Pro B type Natriuretic Peptide
NYHA	New York Heart Association
O2	Oxygen
O2-	Superoxide anion
OD	Omne in die (Once a day)
OMI	Old Myocardial Infarct
PaO2	Partial Pressure of Oxygen
PaCo2	Partial Pressure of Carbon di-oxide
PCV	Packed Cell Volume
PDE-III Inhibitor	Phospho diesterase-III inhibitor
PE	Pericardial Effusion
PET	Positron Emission Ttomography
%	Percent
pH	Potential of Hydrogen
PH	Pulmonary Hypertension
PO	per Os (orally)
PON 1	Paraoxonase 1-
PRC	Post Resuscitation Care

QID	Quater in die (four times a day)
R	Right
RAAS	Renin-Angiotensisn Aldosterone System
RH	Right Heart
RI	Renin Inhibitor
ROS	Reactive oxygen Species
R V	Retro Virus
RV	Right Ventricule
SA -arrest	Sino-atrial Arrest
SA node	Sino-atrial node
SARS-Cov-2	Severe Acute Respiratory syndrome Corona Virus-2
SERCA2a	Sarco Endoplasmic Reticular Calcium 2 + AT pase
SID	Sem el in di e (once a day)
SNS	Sympathetic Nervous System
SPECT	Single Photon Emission Computed Tomography
ST-2	Suppression of Tumorigenicity-2
TEAC	Trolox Equivalent Anti-oxidant Capacity
TMAO	Trimethylamine N-oxide
TI	Tricuspid Insufficiency
TID	Ter in die (three times a day)
TLC	Total Leukocyte Count
TMT	Transient Myocardial Thickening
TNF	Tumour Necrosis Factor
μg	Microgram
U.K.	United kingdom
U.S.A.	United States of America
VADs	Ventricular Assist Devices
VD	Ventro-dorsal
VHS	Vertebral Heart Score
VPC	Ventricular Premature Complex
Vit.	Vitamin
VT	Ventricular tachycardia
X-ray	X-radiation or Rontgen radiation

List of Figures

Section-1
Structure and Functions of the Heart

1

Structure and Function of the Heart

The heart failure is an important cause of morbidity and mortality in animals and man. It is the end stage complex event of cardio-vascular diseases. For having a better understanding of heart failure in animals for adopting rational therapeutic management, it is utmost necessary to have thorough knowledge about heart, its structures, its nerve supplies, its vasculature, current generation and conduction as well as its functioning in the healthy subjects. Further understanding of peculiarities of cardiovascular system in different species of animals and its functioning can facilitate rational approach in the management of heart failure in different species of the animals.

Anatomical Considerations

Cardiac Anatomy

- The heart of the large mammals has many similarities. Nevertheless, the shape, size, and position in thoracic cavity is little different between species (Fig.1).
- The heart is a vital organ located in the thoracic cavity within mediastinum between left and right pleural cavities from about 3rd to 6th intercostal spaces. It is protected by the rib cage.
- As compared to humans, large animals have less pronounced left sided orientation and a more ventrally tilted long axis of the heart (Holt, 1970).
- The size of mammalian heart varies from 0.3 to 1.0% of the body weight. The heart of large animals is elongated with pointed apex.
- The heart of the dog is somewhat ovoid with blunt apex (Fig. 1 A).
- The heart of the cat is elongated and have pointed apex (Fig.1 B).
- Sheep heart is having a blunt apex.
- The heart of the large ruminants (Fig .1C , 1D) lies in the lower anterior part of the thoracic cavity between lungs but projecting more to the left

side to the right. Heart base lies opposite to the 3rd and 6th ribs inclusive and apex lies above the last segment of the sternum.

- Goat heart is comparatively conical (Fig. 1E).
- The apex of pigs' heart is blunt and medially oriented.
- Birds tend to have larger heart (Fig .1F, G) than the mammals. Smaller birds have larger heart than that of the large birds. Avian heart is also four chambered as in case of mammals. The heart is located in the cranial portion of the thoracoabdominal cavity with its long axis slightly to the right. Liver extends caudally from the apex of the heart causing overlapping resulting in an hourglass appearance of the two organs (Fig.1F, G). The heart size in birds is inversely related to a species' body mass. Tricuspid valve is structurally distinct as it consists of single spiral flap of myocardium attached to the free wall of the right ventricle. Mitral valve has three cusps (tricuspid). These valves are connected to Purkinje system. This arrangement is different from that of mammals and allows the leaflets to float up in the atrial chambers. Birds have larger hearts with bigger stroke volumes resulting in larger cardiac outputs than mammals. Birds have lower heart rate per unit size for greater perfusion of tissues of the body. Birds also have greater mean arterial blood pressure as compared to mammals.
- Rabbits' heart (1H) is conical in shape and its cranial border is difficult to distinguish from cranial mediastinum. Heart size is relatively small to body size. Heart is placed high in the chest between the lungs near the sternum. The axis of the heart runs along the ventral chest deviating slightly towards left. The direction of the heart base is towards the front of the thorax .The apex is directed towards the spine and slightly to the left.
- Human heart has a trapezoidal shape with blunt apex. The human heart is somewhat conical and is positioned posteriorly to the body of sternum with one third situated on the right and two third on the left of midline (Fig. 1 I). The heart is anteriorly related to the body of the sternum, and adjoining costal cartilage, left lung and pleura. The heart is posteriorly related to esophagus, descending thoracic aorta, azygos, hemiazygos vein, and thoracic duct. Lungs and pleura are on the lateral side of the heart. Diaphragm is on the inferior side of the heart. Superficially it is related to bifurcation of the main pulmonary trunk. The heart is supplied by left main coronary artery and right coronary artery. Left coronary artery is divided into left anterior descending artery and left circumflex artery.

- The apices of normal dog, pig, sheep and human heart are formed entirely by left ventricle.
- Most of the other quadruped animals have less pronounced left sided orientation and a more ventrally tilted long axis.
- The ratio of heart weight to body weight is different in different species of the animals.
- Sheep and pigs have smaller ratio of heart weight to body weight as compared to dogs.

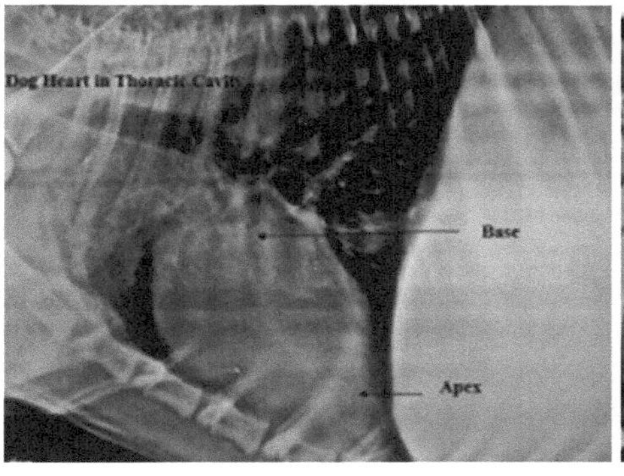

A

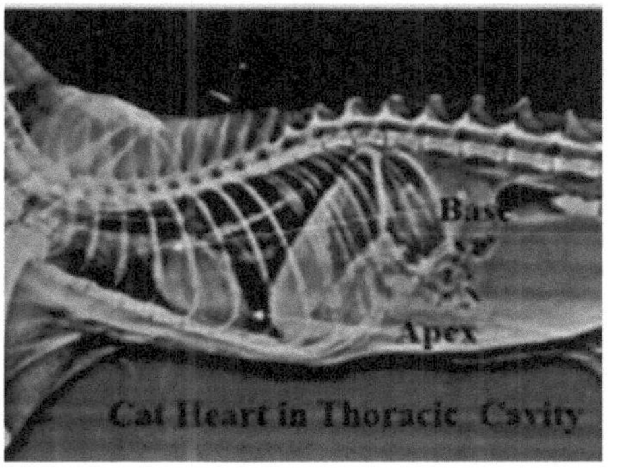

B

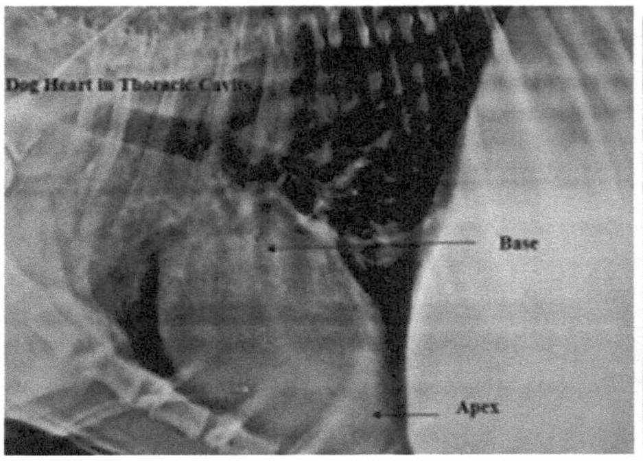

C

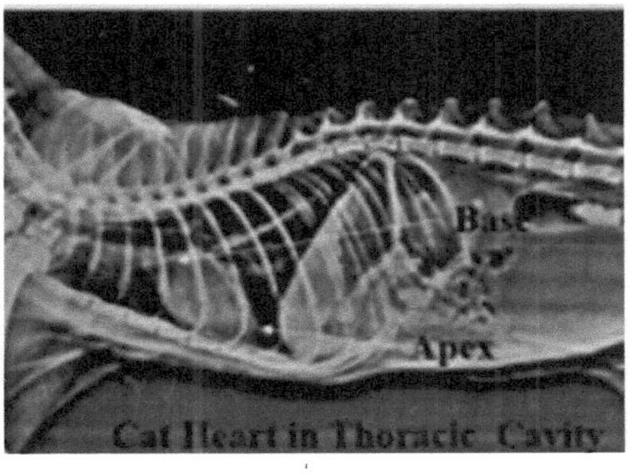

D

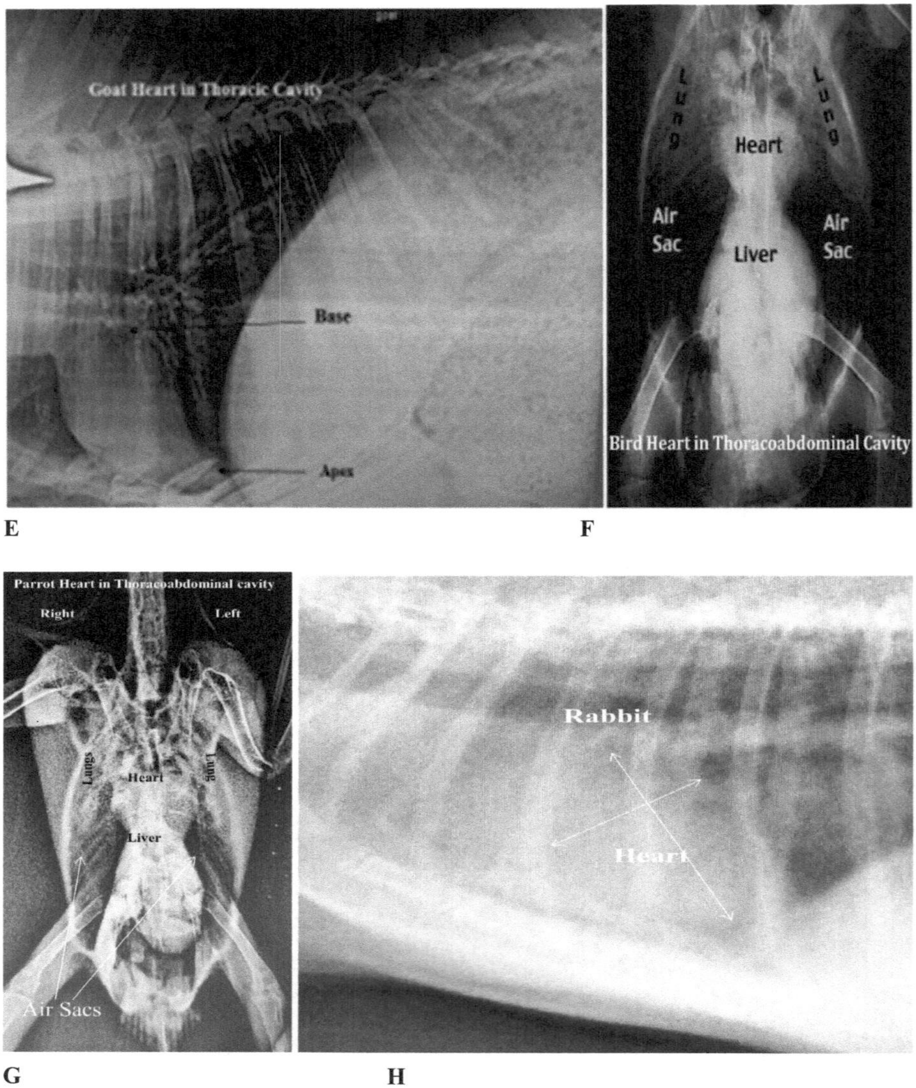
Goat Heart in Thoracic Cavity
Base
Apex
E
Lung
Lung
Heart
Air Sac
Liver
Air Sac
Bird Heart in Thoracoabdominal Cavity
F
Parrot Heart in Thoracoabdominal cavity
Right
Left
Lungs
Heart
Lung
Liver
Air Sacs
G
Rabbit
Heart
H

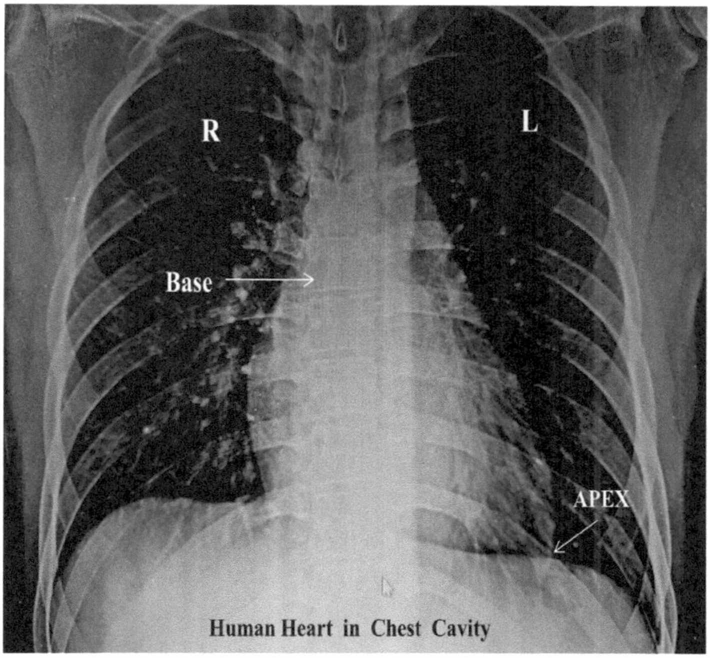

I

Fig. 1: Radiographs of the thorax of different species of animals and human showing shape, size and orientation of the heart of dog **(A)**, cat (B),cow calf **(C)**, buffalo calf **(D)**, goat **(E)**, pet bird **(F)**, parrot **(G)**, rabbit **(H)** and human **(I)**.

- Oscordis (a small fully developed bone lying deep in the atrial septum) is present in the heart of sheep and goats.
- The heart of the fish and reptiles has two and three chambers respectively
- In mammals, humans and birds heart has four chambers (Fig.2).Two upper chambers are left and right atrium and two lower chambers are left and right ventricles
- The heart can broadly be divided as left heart (left atrium and left ventricle) and right heart (right atrium and right ventricle).
- Ventricles have thick muscular wall and left ventricle is larger than the right one. The wall of the right ventricle is thinner than that of left ventricle.
- Both ventricles are separated by interventricular septum.
- Auricles are separated by inter-auricular septum .The wall of the right auricle is thin than that of right ventricle.

- The right auricle and ventricles are separated through tricuspid valve .
- The left auricle and left ventricles are separated through mitral valve
- Cranial and caudal vena cava returns the blood from the distant parts of the body to the right auricle. Pulmonary veins (right and left) brings the oxygenated blood from the lungs to the left auricle.
- Auricles empty their blood into ventricles through respective valves at the junction of auricle and ventricles.
- Ventricles pump their blood out into the blood stream via aortic arches, the brachiocephalic trunk (left ventricle) and the pulmonary trunk (right ventricle) .
- Left ventricular length, thickness of auricles and width of right papillary muscle are larger in sheep than in goats (Mohammadpour and Arbi, 2007).
- The Heart is a muscular organ located in thoracic cavity. It is a very vital organ that keeps on pumping the blood through the blood vessels since birth to death.
- There are three types of cardiac muscle viz. atrial, ventricular and noncontractile.
- Heart wall is made up of epicardium (outermost layer), myocardium (muscle layer) and endocardium (inner most layer lining the heart chambers and covering the cardiac valves).
- Cardiac muscles have special properties such as automaticity, excitability, refractoriness, conductivity and contractility (Berne and Levy, 1977, Fozzard, 1977).Cardiac muscle fibers are capable of initiating their own impulse (automaticity). When resting potential is reduced to a certain critical level, cardiac muscle cell is excited. During contraction, cardiac muscle cells will not respond to the external stimuli. This phenomenon is termed as refractoriness. Activation of an individual cardiac muscle cell stimulates the neighboring cardiac muscle cell. This phenomenon is called conductivity. In response to electrical current cardiac muscle contracts. This property of cardiac muscle cell is called contractility.
- Cardiac muscle fibers are shorter in length but larger in diameter than skeletal muscles fibers and individual fiber is Y shaped. Cardiac muscle fibers have abundant sarcoplasm and large numerous mitochondria. Specialty of cardiac muscles is that they do not fatigue, cannot be repaired when damaged and regulated by autonomic nervous system.

- The wall of the heart is thick consisting of three layers viz. epicardium, myocardium and endocardium.
- The heart is enclosed in a protective sac known as pericardial sac having a small amount of pericardial fluid. The pericardium is fixed to great arteries at the base of the heart and attached to sternum and diaphragm. Pericardium consists of three layers (the serous visceral pericardium also known as epicardium, the serous parietal pericardium, and fibrous pericardium).
- Both atria are located at the base of the heart. Base of the heart receives all great vessels.
- Atria and ventricles are separated by atrioventricular valves (mitral valve between left atrium and left ventricle; tricuspid valve between right atrium and right ventricle) connected to papillary muscles through chordae tendinae.
- Tricuspid valve of dogs has two leaflets while tricuspid valve of swine has three leaflets.
- Characteristics of porcine mitral valve are similar to humans.

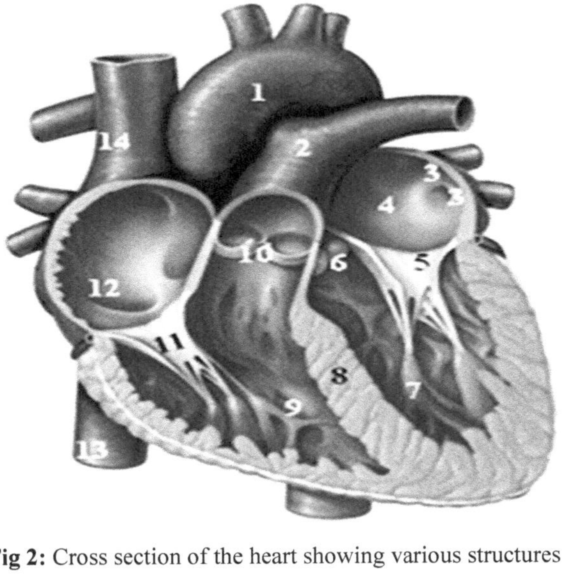

Fig 2: Cross section of the heart showing various structures

Conduction system

- Almost all mammalian heart have similar conduction system. Conduction system of the heart is shown in Fig. 3 A and B .

- Sinoatrial node (SA node), interatrial bundle, atrioventricular node (AV node), bundle of His, right and left bundle branches, and Purkinje fibers are the main component of the conduction system.
- SA node is the normal pace maker that is situated high on the right atrial wall near the junction of superior vena cava and right atrium. SA node is the seat of current generation. The SA node has unique three dimensional heterogeneous structure characterized by small pace maker myocytes enmeshed within fibrotic strands.
- Current flows from SA node to ventricles through AV node, bundle of His, bundle branches and Purkinje fibers. The speed of current flow from one point to other is not uniform.
- The wave of depolarizing current spreads from SA node to across atrium and to AV node.
- The current from AV node spreads to bundle of His.
- The current from bundle of His spreads to left and right bundle branches.
- The current from right bundle branch spreads to right ventricle.
- The current from left bundle branch spreads to Purkinje fibers through anterior and posterior fascicle.
- The Purkinje fibers supply the current to ventricular myocardium.

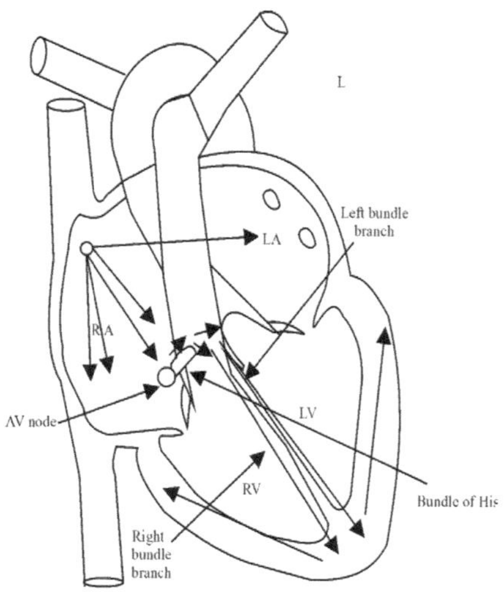

Cardiac Conduction system in Dogs Diagrammatic sketch

A

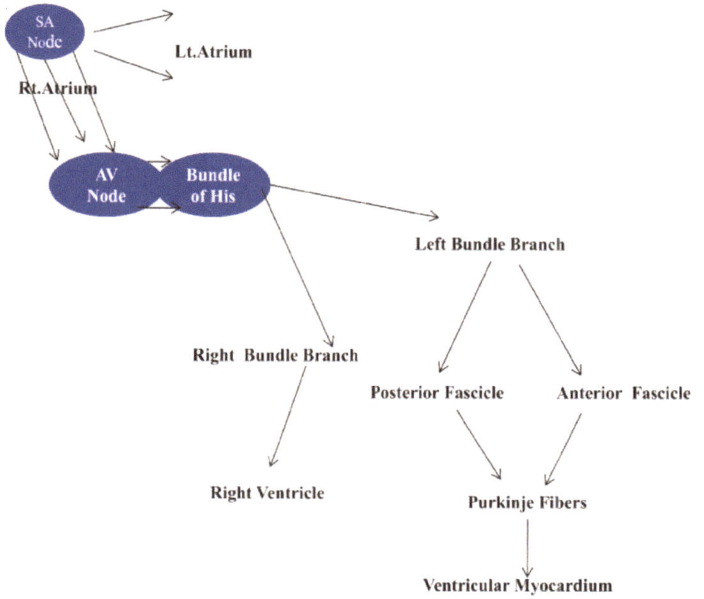

Diagrammatic representation of current generation and its flow in the heart of the dog

B

Fig. 3 (A,B): Diagrammatic sketch of current flow pattern in the heart of a dog

Nerve Supply to the Heart

- Autonomic nervous system consists of sympathetic and parasympathetic nerves. Autonomic nervous system influences heart rate, conduction system and contractility of atria and ventricles. The heart is generously innervated with both sympathetic and parasympathetic nerves. Nerve impulse transmission in both parasympathetic and sympathetic nerves is chemically mediated.
- SA node, AV node and atrial muscle fibers (to some extent) are innervated by parasympathetic nerves. Ventricles have few inhibitory fibers (parasympathetic) than the excitatory fibers (sympathetic) .The vagus nerve supplies parasympathetic innervations. Parasympathetic nerves originates in the medulla oblongata. Right vagus nerve fibers predominantly affect the SA node. AV conduction tissue is predominantly affected by the left vagus nerve fibers. Vagus stimulation increases parasympathetic tone resulting into decrease in heart rate and slow

conduction through AV node. Excessive vagal stimulation can adversely affect the rhythmicity of Sino atrial node.

- The sympathetic nerves richly supply ventricles. Sympathetic nerve stimulation increases heart rate (positive chronotropic effect) and contraction force (positive inotropic effect). Right and left sympathetic nerves to ventricles differ in their effect in dogs and cats (D'Agrosa, 1977, Schwaryz *et al.*, 1977). Left sympathetic nerve is more dominant and its fibers supply to posterior ventricular wall and right sympathetic nerve fibers supply mainly to anterior ventricular wall (Lown *et al.*1977). Stimulation of left sympathetic nerves causes AV junctional and/ or ventricular arrhythmias, prolongation of Q-T interval and T wave changes (D'Agrosa, 1977).
- Cardiac apex has abundant number of efferent nerve fibers.
- Sensatory nerve fibers are found mainly in auricles and rarely in ventricles.
- Canine's AV node is innervated by many excitatory and inhibitory fibers and also by considerable number of sensatory fibers.

Peculiar Characteristics of the Heart of Different Species

Dog

- Pericardium is attached to the sternal part of the diaphragm by pericardiacophreic ligament.
- Apex is blunt or rounded.
- Heart is obliquely placed at the level between 3rd to 6th intercostal space.
- Right and left longitudinal grooves meet together to the right of the apex.

Cats

- Orientation of the long axis of cardiac silhouette is more oblique.

Horse

- Heart is located at the level of 2nd to 6th intercostal space.
- Long axis of cardiac silhouette is oriented vertically.
- Heart base is broader.

- Surfaces are flatter.
- Cranial border of the heart is more convex.
- Caudal border of the heart is nearly vertical.
- Heart apex is placed at 7th sternebra.
- No meeting of left and right longitudinal grooves.
- Deep penetration of Purkinje fibers.

Rabbit

- The heart is conical.
- Cranial border of the heart is difficult to distinguish from cranial mediastinum in lateral view (Pariaut, 2009).
- Heart size is relatively small to body size.
- The heart is placed high in the chest between the lungs near the sternum.
- Pericardium is attached to the diaphragm keeping the heart and major blood vessels in position in the thoracic cavity.
- The axis of the heart runs along the ventral chest deviating slightly towards left.
- The direction of the heart base is towards the front of the thorax .The apex is directed towards the spine and slightly to the left.
- The pulmonary artery and its branches are very muscular.
- The coronary arteries are less in number.
- Myocardial ischemia may develop when compressed due to poor collateral blood circulation.
- Aorta has pressure sensitive receptors.
- The sinus venosus and truncus arteriosus are not well marked.
- Left subclavian artery arises separately.
- Aortic arch is strongly curved.

Ruminants

- Long axis of cardiac silhouette is oriented almost vertically.
- Oscordis, two small triangular bones lodged at the base of ventricles on either side of aortic ring, are characteristic of the heart of domestic ruminants.
- Purkinje fibers are deep penetrated.

Pet Birds

- The heart is four chambered.
- The heart is larger relative to body size and mass.
- The mitral valve has three cusps.
- The tricuspid valve has single spiral flap of myocardium.

Man

- Human heart has trapezoidal shape with blunt apex.
- Two third heart is on the left of midline.

Physiological Considerations

Functions of the Heart

- One of the primary functions of the heart is to pump blood throughout the body. The heart is a unique astonishing pump performing a huge task in extremely variable conditions over the life time of an individual.
- The heart delivers oxygen, hormones, nutrients and other components to various parts of the body, including the heart itself through its pumping action.
- The heart receives back deoxygenated blood and blood carrying metabolic waste products from the body and pumps it to lungs for oxygenation.
- The heart also ensures that adequate blood pressure is maintained in the body.
- Although heart functions independently, its performance is under the influence of humoral and neural factors.
- The function of the pericardium is to limit distension of the heart during filling.

- Right atrium receives the blood returning to heart via cranial and caudal vena cava.
- Right ventricle receives all the blood from right atrium and pump it to lungs for gaseous exchange.
- Left atrium receives the blood returning from pulmonary veins.
- Left ventricle receives all the blood from the left atrium and pump it through systemic arterial tree to different organs.
- A-V valves allow blood flow in unidirection from atria to ventricles during ventricular diastole when ventricular pressure falls below the pressure of the atria. Three phases of blood flow across the A-V valves occur as rapid filling (at the beginning of ventricular diastole), diastasis (minimum blood flow at the stage) and atrial contraction at the end of ventricular diastole.
- Semilunar valves permit blood flow from ventricles into the arterial trees during ventricular systole.

Blood flow through Heart

Blood flow through heart and lung is in four steps as follows;

- Deoxygenated blood from body is received in right atrium. It is pumped into right ventricle through tricuspid valve during atrial systole.
- The right ventricle pumps the deoxygenated blood to the lungs through the pulmonary valve during ventricular systole.
- Oxygenated blood from lungs is received in left atrium. It is pumped into left ventricle through the mitral valve during atrial systole.
- The left ventricle pumps the oxygenated blood to the rest of the body through the aortic valve during ventricular systole.

Cardiac Cycle

- One cardiac cycle comprised of systole (contraction) and diastole (relaxation) of atrium and ventricles and valve movements.
- The cycle repeats with every heartbeat. It is systole and diastole.
- Systole causes isovolumetric contraction resulting into ejection of blood.
- While diastole causes isovolumetric relaxation and filling of chambers.

- Right and left ventricular cycles are identical except for the peak pressures.
- The repeated systole and diastole of the heart is termed as cardiac cycle. It can be divided into three stages.
- In electrocardiogram these reference points are 'P' wave (atrial depolarization), 'QRS complex (ventricular depolarization) and 'T' wave (Ventricular repolarization). Thus one complete cardiac cycle is represented in the electrocardiogram by 'P' wave, 'P-R interval' 'QRS' complex, 'S-T segment' and 'T' wave (Fig .4). After 'P' wave atria will contract and after 'QRS' complex ventricles will contract.

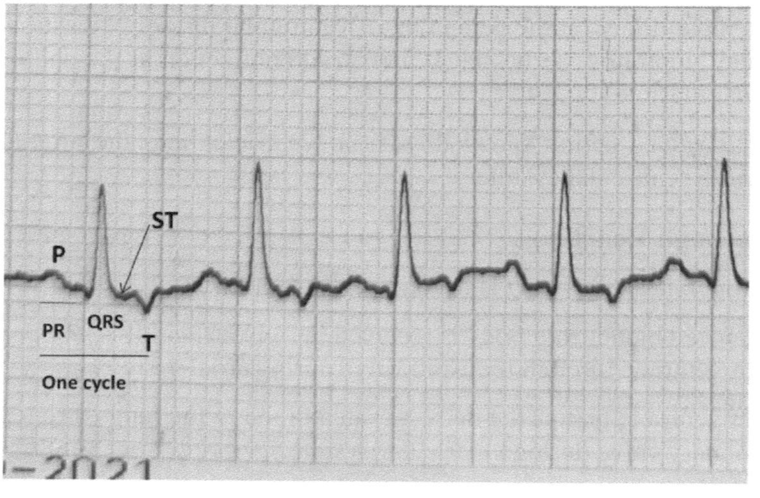

Fig 4: Electrocardiogram of a dog (sensitivity 1.0, speed 25 mm/sec) showing components of one cardiac cycle.

Influence of Hormones, Oxygen, Carbon dioxide and pH on the Heart

- Hormones such as epinephrine, nor-epinephrine, acetylcholine, glucagon, insulin, mineralocorticoids, adrenocorticotropic hormone (ACTH), vasopressin, oxytocin, angiotensin I, angiotensin II, thyroxin and prostaglandins have direct influence on the heart.
- A histamine receptor system analogous to the adrenergic receptor system has been reported in the heart (Kline *et al*., 2015). But physiological role of histamine in heart regulation is not fully established.
- Oxygen, carbon dioxide and pH also influences cardiorespiratory system via actions on carotid body and central chemoreceptors. Hypoxia and hypercapnia can depress cardiac contractility.

References

Berne, R.M. and Levy, M.N. (1977). Cardiovascular Physiology. 3rd edn., St. Lcuis, C V. Mosby.

D'Agrosa, L.S.(1977).Cardiac arrhythmias of sympathetic origin in the dog. Am. J. Physiol. 233: H535-540.

Fozzard, H.A. (1977). Cardiac muscle: excitability and passive electrical properties. Prog. Cardiovasc. Dis. 19: 343-349.

Holt, J.P. (1970) . The normal pericardium. Am. J. Cardiol. 26:455-465.

Kline, D.D., Hasser, E.M., and Heesch, C.M. (2015). Regulation of the Heart. In:Dukes' Physiology of Domestic Animal. Reece, W.M., Erickson ,H.H., Goff, J.P. and Uemura, E.E. (eds). Willey Blackwell. pp 341-351.

Lown, B., Verrier, R.L. and Robinowitz, S.H. (1977). Neural and psychologic mechanisms and the problem of sudden cardiac death. Am.J.Cardiol. 39:890-902.

Mohammadpour, A. and Arabi, M. (2007). Morphological study of the heart and oscordis in sheep and goat. Indian Vet.J.84:284-287.

Pariaut, R.(2009). Cardiovascular Physiology and Diseases of the Rabbit. Vet. Clinics North Am. Exotic Anim. Pract. 12 : 135-144.

Schwartz, P.J., Verrier, R.L. and Lown, B. (1977). Effect of stellectomy and vagotomy on ventricular refractoriness in dogs. Cir. Res. 40: 536-540.

Section 2
Heart Failure in Canine

2

Heart Failure in Canine

Heart failure in canine is being recognized with an increasing frequency throughout the world. It is the end stage pathological event progressively occurring in heart diseases. Heart failure is the condition in which heart is unable to meet the metabolic and nutritional demands of the tissues of the body due to insufficient cardiac output. Heart failure has also been observed with normal cardiac output in conditions like hyperthyroidism or anemia where tissue metabolic needs are increased. All heart diseases may not lead to heart failure, but heart failure is mostly due to heart diseases. The words "heart disease" and "heart failure" are not the same. Many heart diseases may not require therapeutic intervention but therapeutic intervention is necessary in cases of heart failure.

Prevalence of Heart Failure

Heart failure, a syndrome resulting from heart disease, is an important cause of morbidity and mortality in dogs and cats. There are wide variations (10 to 26 %) in the prevalence rate of heart diseases in canine in different countries. Estimates indicate that two million dogs suffer from heart diseases annually in U.K.; and 10% of dogs suffer from heart diseases in USA. Heart failure occurs as peripheral circulatory failure and /or congestive heart failure. Of these congestive heart failure is more commonly observed in dogs and cats. In dogs valvular heart disease and cardiomyopathy are the important cause of heart failure. Prevalence of congestive cardiomyopathies varies from 0.3 % in Cocker Spaniels, 0.7 % in English Cocker Spaniel (Sisson and Thomas, 1995), 3.4 % in Boxer, 6.0 % in Scottish Dachshund (Sisson and Thomas, 1995) to 43.7 % in Doberman (O' Grady and Horne, 1995). Calvert *et al.* (1997) observed that 25 % of the affected Doberman died suddenly prior to onset of overt signs of congestive heart failure. Prevalence of congestive heart failure has been observed as 19 % (57/300) in cases of cardiopathies in dogs at this hospital (Varshney *et al.*, 2011).

Etiology of Heart failure

There are many etiological factors of heart failure varying from severe arrhythmias, structural and functional defects, nutritional deficiencies, infections, shunting, pericardial effusions to insufficient preload (Fig. 5)

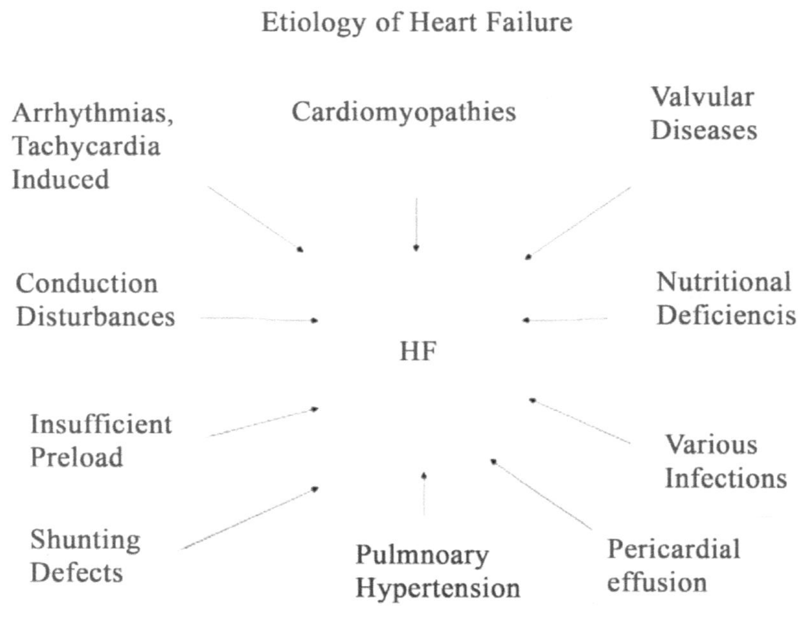

Fig 5: Showing various etiological factors leading to heart failure in dogs

- Cardiomyopathy (Fig.6) is an important cause of heart failure in large breeds of the dog. It is of two types i.e. primary and secondary. Primary Cardiomyopathy is also known as idiopathic cardiomyopathy and is classified as dilated cardiomyopathy, hypertrophic cardiomyopathy and restrictive cardiomyopathy. Of these dilated cardiomyopathy is more prevalent in dogs. It was first reported in 1970. It is primarily a systolic dysfunction. Hypertrophic cardiomyopathy is less common in dogs. But it is more prevalent in cats. There is concentric hypertrophy of the left ventricle and it is a diastolic dysfunction. Restrictive cardiomyopathy is seen in cats. It is characterized by fibrous infiltration of ventricular wall and endocardium. It is also a diastolic dysfunction.

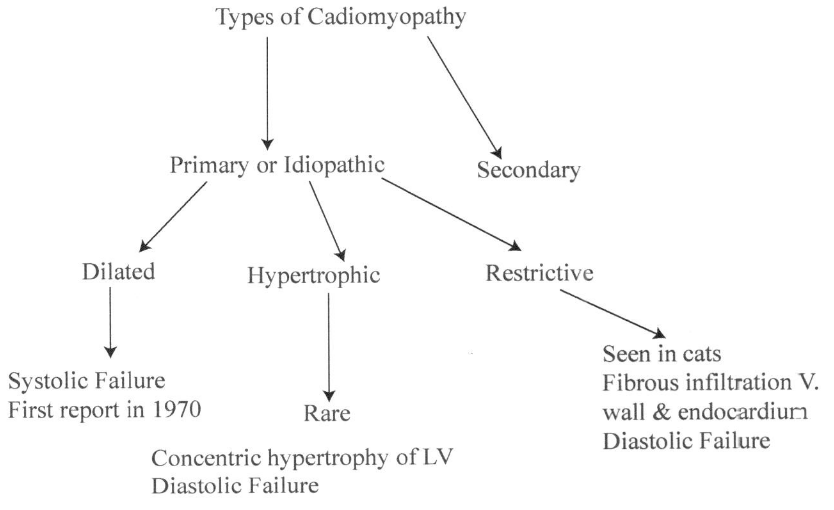

Fig. 6: Showing primary or idiopathic types of cardiomyopathies

Secondary cardiomyopathies are due to other cardiac reasons as well as non-cardiac reasons as shown in the Fig.7. Reasons for secondary cardiomyopathies are ischemia due to microscopic intramural myocardial infarction (MIMI), arterial hypertension; endocarditis; gastric dilatation and volvulus; and infections caused by viruses, bacteria, protozoa and rickettsia; physical factors like hyperthermia, traumas; metabolic diseases (diabetes mellitus, hypo or hyperthyroidism, uremia, electrolyte imbalances, hyperadrenocorticism); nutritional imbalances (taurine, L-carnitine, Vitamin E, Vitamin B1, anemia), drug toxicity (doxorubicin), immunological disease (Lupus erythromatous – LE) and other diseases.

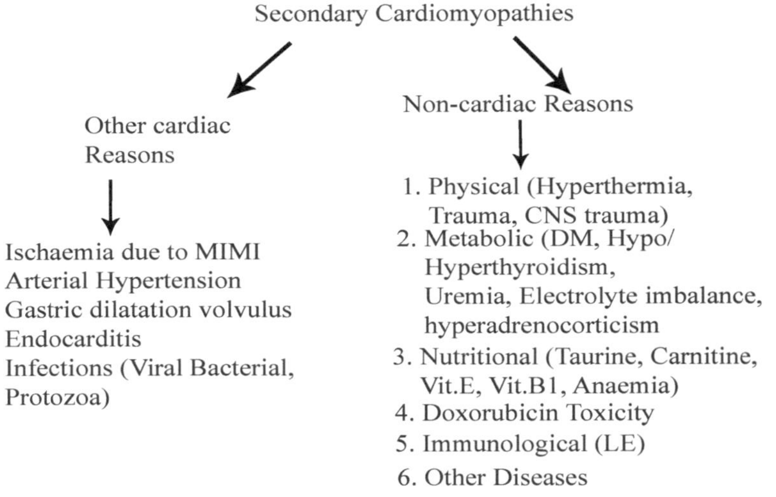

Fig. 7: Showing different reasons of secondary cardiomyopathies

- Arrhythmia such as severe bradycardia, tachycardia, sick sinus syndrome, atrial tachycardia, ventricular tachycardia (Fig.8) may also lead to heart failure.

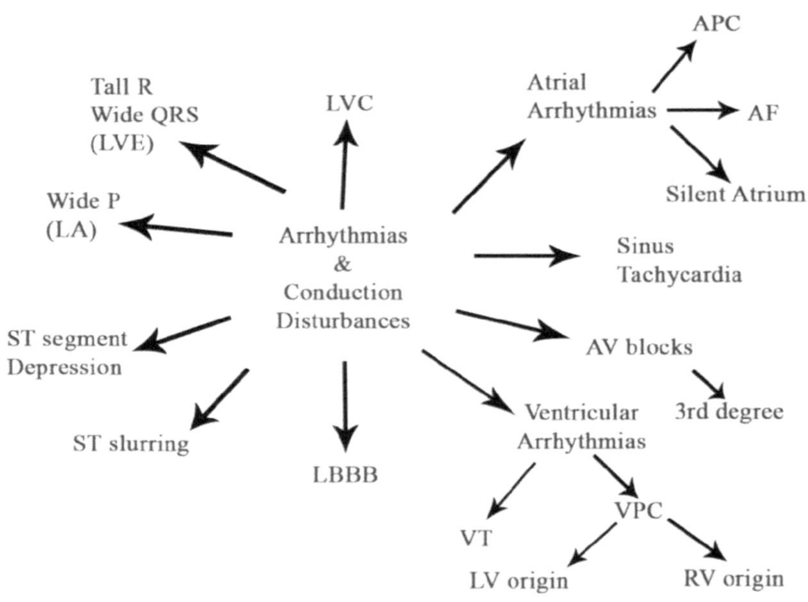

Fig. 8: Showing different arrhythmias and conduction disturbances in dogs.

- Conduction disturbances such as severe 3rd degree heart block (Fig.8).
- Valvular diseases such as mitral and tricuspid valve insufficiencies are the common cause of heart failure in small breeds of dogs.
- Loss of myocardial contractility owing to taurine deficiency.
- Tachycardia induced heart failure has also been recorded.
- Impaired relaxation as seen in concentric hypertrophy or myocardial infiltrate.
- Various cardiac infections (bacterial, viral, protozoan, rickettsia, fungal or heart worm) may lead to heart failure . Recently Romito *et al.* (2021) have reported a case of myocardial injury complicated by systolic dysfunction in a Covid-19 positive dog in Italy. Another report from U.K. has described, for the first time, Covid-19 B.1.1.7 variant infection in cats and dogs with congestive heart failure due to myocarditis (Ferasin *et al.*, 2021).
- Right heart failure in a Pomeranian dog suffering from fatal dirofilariasis has recently been reported from this hospital (Varshney, 2018).
- Shunting defects may also lead to cardiac failure.
- Insufficient preload as seen in blood loss, pericardial effusions, peripheral vasodilation or pulmonary hypertension.

Types of Heart Failure

There are two types of heart failure (Fig. 9)

- Forward Failure (cardiac output failure): Forward failure is due to inadequate cardiac output and is characterized by marked weakness, exercise intolerance, pale mucus membrane, and cold extremities.
- Backward Failure– It is also known as congestive heart failure and occurs due to increased end diastolic pressure of left ventricle owing to volume overload leading to fluid buildup.
    - Backward failure can be further grouped as left heart failure and right heart failure.
    - In case of left heart failure pulmonary venous congestion followed by pulmonary edema develops leading to pleural and abdominal effusions in later stage.

- In case of right heart failure, peritoneal effusions (ascites) and /or pleural effusion develops due to increase in mean capillary hydraulic pressure resulting into an increase in the fluid transfer rate from capillaries to peritoneal space.
- Many cases of heart failure may not have congestive heart failure.

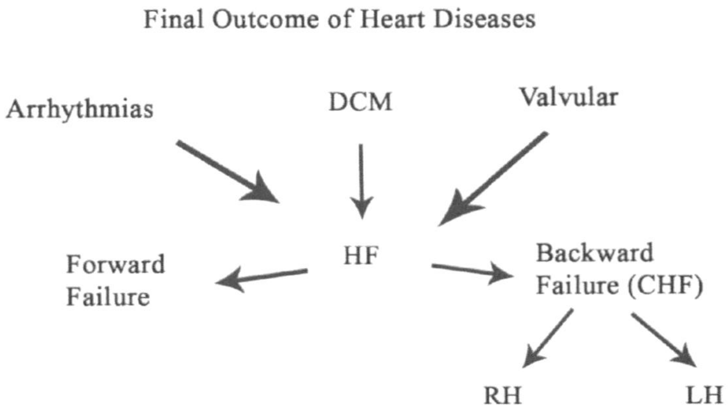

Fig. 9: Showing final outcome of heart diseases and types of heart failure

Compensatory Mechanism in Heart failure

In response to failing heart, many mechanism start operating to compensate the changed requirements of the body.

a. Neuro-hormonal Compensatory Mechanism

- The most accepted compensatory mechanism is neuro-hormonal (Fig.10). With the failing heart cardiac output is decreased leading to fall in blood pressure. The fall in blood pressure causes decreased renal perfusion and escalates release of vasopressin. The release of vasopressin causes vasoconstriction and increased water resorption. Increased vasoconstriction in turn causes increase in afterload. The decreased renal perfusion also causes efferent arteriolar constriction and renin activation. Efferent arteriolar constriction causes further increase in sodium resorption resulting in to increased water resorption and increased venous pressure. The renin activation, due to decreased renal perfusion, increases angiotensin I (AT-I) that is converted to angiotensin II (AT-II) with the help of angiotensin converting enzyme (ACE). The angiotensin II (AT-II) increases water resoption and aldosterone release. The increased water resorption leads to increased venous pressure that in turn causes increase in preload, edema and effusions. Increase in aldosterone also causes increased sodium resorption leading to increased venous pressure and eventually increase in preload, edema and effusions.

Diagrammatic Representation of Events in Response to Heart failure

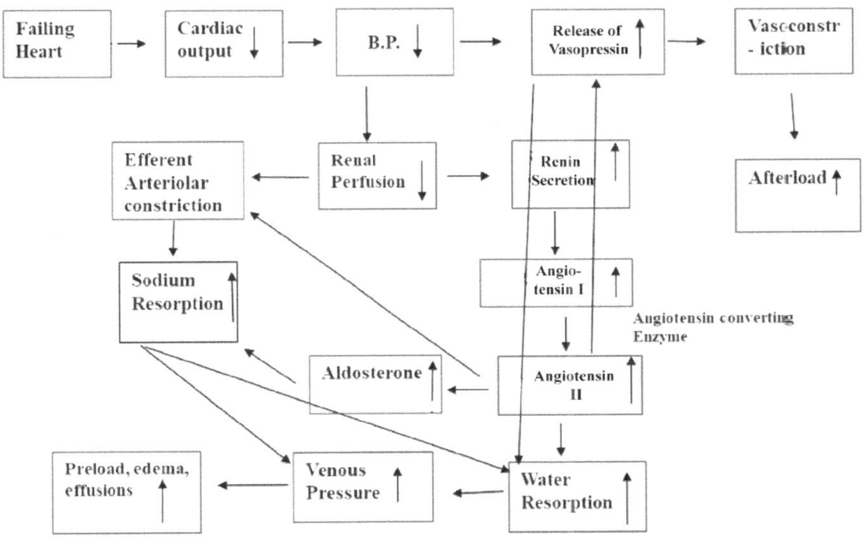

Fig. 10: Diagrammatic representation of compensatory neuro-hormonal events in heart failure.

b. Other Compensatory Mechanisms

Apart from neuro-hormonal mechanism, some other mechanisms also operate to compensate the changed requirements of the body when heart is compromised.

- Increase in sympathetic tone- It causes an increase in heart rate, vasomotor tone, and venomotor tone, release of rennin from kidney, increase in contractility, promoting arrhythmias and increasing in myocardial oxygen demand.
- Redistribution of Cardiac Output - Cardiac output is redistributed .First priority is to maintain blood supply to coronary and cerebral vascular beds. In this effort blood supply to cutaneous vascular bed, splanchnic vascular bed and renal vascular beds is reduced resulting into pale

mucus membrane, slow capillary refill time, cool extremities, impaired intestinal absorption, bowel ulcer and hemorrhages due to bowel ischemia, reduced glomerular filtration, and retention of sodium and water.

- Increased thirst- Increased water intake due to increased thirst can result into fluid overload.
- Increase in arginine vasopressin (ADH)- Increase in arginine vasopressin enhances arterial vasomotor tone and fluid retention resulting into an increase in afterload and preload.
- Arterial vasoconstriction- Arterial vasoconstriction is increased due to elevation in angiotensin II, arginine vasopressin and endothelin level. Due to vasoconstriction blood pressure is maintained but afterload is increased aggravating the heart failure.
- Arterial vasodilation- Arterial vasodilatory prostaglandins, atrial natriuretic peptide, endothelium derived relaxation factors are responsible for arterial vasodilation. These factors becomes blunted in heart failure. Factors responsible for arterial vasoconstriction dominate the scene promoting the persistent progression of heart failure.

In long term these compensatory mechanisms usually have deleterious effect on the heart leading to progress in heart failure.

c. **Other Changes**

- Recent studies have shown that besides neurohormonal compensatory mechanism many other changes such as changes in ions, immunological indices, oxidative stress indices, vitamin levels, hepatic and renal indices, platelets, myocardial structure, blood pressure, and gut micro flora and peristalsis are occurring in the body that influence the outcome of cases with heart failure (Fig.11.).

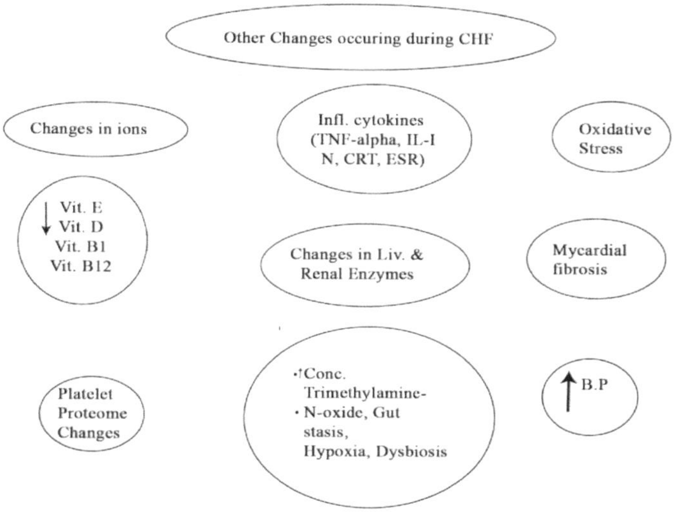

Fig. 11: Showing other changes occurring during the process of congestive heart failure.

- Presently therapeutic interventions undertaken in the management of heart failure in dogs provide short term success as dogs suffering from heart failure do not survive long (more than 9 to 12 months) despite best efforts. Most accepted dominant neurohumoral/neurohormonal theory, based on malfunctioning of sympathoadrenal and renin–angiotensin-aldosterone system (RAAS) fails to explain heart failure syndrome completely (Fonfara *et al.*,2014; Janus *et al.*,2017).Though, dogs with heart failure owing to dilated cardiomyopathy are being treated with effective drugs such as beta- blockers, ACE-inhibitors, calcium sensitizers, and phosphodiesterase inhibitors, the course of the disease remains progressive. This fact has led to speculate that in addition to neurohumoral mechanism some other systems must be involved in the pathogenesis of heart failure in humans and animals. Changes observed in different immunological and oxidative stress indices in different studies in dogs and humans with heart failure are represented in Fig. 12 .

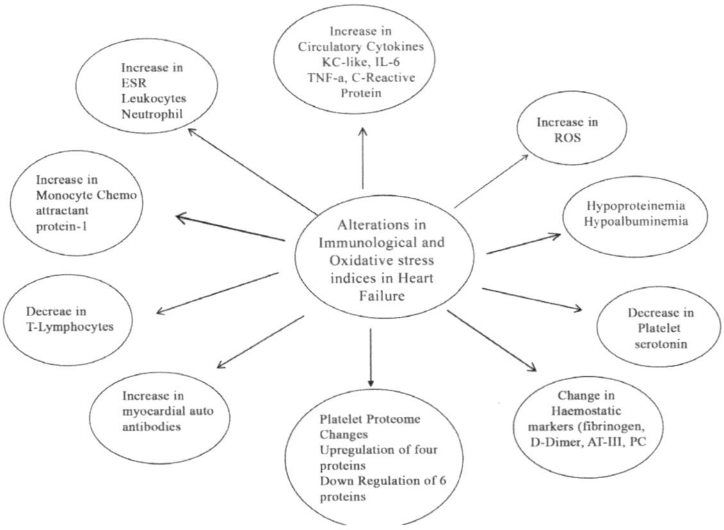

Fig. 12: Changes in Immunological and Oxidative Stress indices in dogs with heart failure

- Although, dilated cardiomyopathy (DCM) and myxomatous mitral valve degeneration disease (MMVD) are commonly considered non-inflammatory conditions (Reimann *et al.*, 2016, Schultheiss *et al.*, 2019), studies have shown increased levels of circulating inflammatory cytokines in dogs and humans with congestive heart failure due to MMVD (Zois *et al.*, 2012, Fonfara *et al.*, 2013) and DCM (Reimann *et al.*,2016, Vatnikov *et a*l.,2019).During recent years immunological concept of the pathogenesis of chronic heart failure is gaining prominence on the basis of elevation in the values of erythrocyte sedimentation rate, leukocytes, neutrophil numbers, C-reactive protein and pro-inflammatory cytokine in humans with heart disease suggesting chronic aseptic inflammation (Cunningham *et al.*, 2012, Zavalishina *et al.*, 2016). Rubio *et al.* (2020) observed an increase in positive acute phase proteins and selected inflammatory cytokines such as KC like (Keratinocyte Chemotactic-like) and decrease in anti-oxidant biomarkers indicating that inflammation and oxidative stress acted as collaborative partners in the pathogenesis of heart failure and these biomarkers could be used to monitor the severity of the disease and also the effect of treatment. They did not observed significant differences in IL2, IL6, IL7, IL8, IL10, IL15, IL18, IP10, MCP-1 (Monocyte Chemotactic Protein-1),GM-CSF (Granulocyte Macrophase- Colony Stimulating Factor),TNF-alpha (Tumour necrosis factor-alpha), and IFN-Y (Interferon-gamma) in dogs of different stages of heart failure. Increased expression and release of inflammatory cytokines such as tumour necrosis factor(TNF-alpha) and serum C-reactive protein(CRP) has also been reported in humans (Dekker et al., 2014) and dogs with congestive heart failure (Domanjko et al.,2018).

- Cell oxygenation and cellular metabolism is impaired in dogs with heart failure resulting into production of large amount of free radicals leading to oxidative stress. In oxidative stress, the production of reactive oxygen species (ROS) is in excess relative to anti-oxidant defense. The increase in free radicals causes cellular damage. In heart failure, the main sources of free radicals are mitochondria (Ide *et al.*, 1999), Xanthine oxidases and NADPH oxidases (Byrne *et al.*, 2003) and uncoupled nitric oxide synthase. Chronic increases in ROS production in mitochondria in failing heart leads to damage of mitochondrial DNA, decline in cell function, further generation of ROS and cellular injury, impaired contractile function, mediate apoptosis , stimulate cardiac fibroblast proliferation and activate the matrix metalloproteinase. Studies conducted by Rubio *et al.* (2020) revealed lower values of all anti-oxidant biomarkers (Paraoxonase1-PON1, Trolox equivalent anti-oxidant capacity-TEAC, Cupric reducing anti-oxidant capacity-CUPRAC and thiol) in dogs with severe congestive heart failure. The decrease in anti-oxidant biomarkers may be connected with the increase in inflammatory biomarkers as imbalance between antioxidants and oxidants can escalate inflammatory cytokines (Mak and Newton, 2001).
- Increase in monocyte chemo attractant protein 1 (MCP1) and decrease in interleukins have also been reported in dogs with congestive heart failure (Zois *et al.,* 2012). But further confirmatory studies are needed to explain how these biomarkers can change in severe cases.
- Besides the toxic effect of the products of chronic aseptic inflammation, stasis and hypoxia of the gut results into malabsorption, deranged tissue metabolism, and decreased liver detoxification eventually leading to endogenous intoxication and secondary damage to other organs (Petric *et al.*, 2018).
- Vatnikov *et al.* (2019) observed leukocytosis, neutrophilia with shift to the left, T-lymphocytopenia, an increase in circulating immune complex, an increase in myocardial autoantibodies, hypoproteinemia with hypoalbuminemia, deranged liver and kidney functions in dogs with heart failure due to dilated cardiomyopathy.

- Platelets play an important role in the genesis of heart diseases and its proteins have been involved in the pathophysiology of heart failure in humans. Since platelets contain about 5000 proteins, their role in the progression of heart disease may be higher than expected. During the progression of congestive heart failure in cases of myxomatous mitral valve disease, turbulent high velocity blood flow and changes in blood shear stress around degenerative mitral valve leaflet may produce platelet activation in dogs (Moesgaard *et al.,* 2009) leading to sudden death by development of intra-myocardial coronary arteriosclerosis and micro thrombi (Moesgaard *et al*., 2009). Levent *et al.* (2020) identified 104 common differentially expressed proteins on platelets. Out of these, four proteins (Guanine-nucleotide binding protein, apolipoproteins (A-II and C-III) and clusterin) were up regulated and 6 (CXC-motif-chemokine-10, cytochrome- C-oxidase subunit-2, cathepsin-D, serine/threonine- protein- phosphatase –PP1-gamma-catlytic subunit, creatinine- kinase- B type , and myotrophin) were down regulated in dogs with congestive heart failure. These proteins are associated with several molecular functions, biological processes, signaling system, and immune inflammatory responses and may have role in pathophysiology of congestive heart failure due to myxomatous mitral valve disease.
- A decrease in platelet serotonin concentration has been reported in degenerative mitral valve disease in dogs with pulmonary hypertension. However, the plasma serotonin concentration was not different between normal and DMVD dogs with or without pulmonary hypertension (Tangmahakul *et al*., 2021). Serotonin, a biogenic amine, is involved in dysfunctions of organs including cardiovascular system. It has been reported that serotonin in platelets may play an important role in the pathogenesis of DMVD in dogs (Cremer *et al*., 2014).
- Homeostasis disturbance is the common complication of congestive heart failure in human. Heart failure is associated with an increased risk of thromboembolism, stroke and sudden death in humans. Platelet abnormalities have been described in cases of heart failure. A study on hemostatic markers in dogs with congestive heart failure due to mitral valve disease revealed significant changes in prothrombin time (PT),fibrinogen, D-Dimer, antithrombinIII (AT III),protein C (PC), Factor VII (F-VII),factor –VIII (F-VIII), hematocrit (HCT), platelet number (PLT), total plasma proteins (TP), and albumin (ALB) and the most important determinants of prothrmbotic state were increased fibrinogen, and D-dimer concentrations and decreased antithrombin III (AT-III) and protein C (PC) activities (Prihirunkit *et al* .,2014).

- Cortisol is secreted in physiological, pathological stress or inflammatory conditions. Its effect in body is mediated through binding to intracellular glucocorticoid receptors (GRs). Cortisol binds with mineralocorticoid receptors (MRs) with same affinity as aldosterone. In healthy subjects, it occupies the MR binding site without activating the receptor. In inflamed or hypoxic tissues, cortisol is able to activate MRs and mimic the actions of aldosterone (Dooley *et al.*, 2012). High levels of cortisol and aldosterone are associated with the worst prognosis in humans with chronic congestive heart failure (Güder *et al.*, 2015). It appears that the benefit of blocking MRs may have more to do with blocking cortisol than with blocking aldosterone. Studies have shown prognostic value of cortisol in humans with congestive heart failure. The effect of endogenous cortisol levels in congestive heart failure in dogs is not clear and remains to be further investigated.

Clinical Aspects of Heart Failure

History- Owners generally report following observations in dogs affected with heart failure.

- Lethargy/syncope- This is related to decrease cerebral perfusion due to reduced cardiac output.
- Weakness/fatigue- This is related to decrease skeletal muscle perfusion due to reduced cardiac output.
- Respiratory distress signs- Dyspnea/ orthopnea/ wheezing/coughing are the signs generally narrated by the owners. These signs are related to pulmonary edema or pleural effusion. In many cases history of coughing is not available.
- Abdominal distension- It may be due to ascites, hepatomegaly or splenomegaly.

Clinical Signs -The clinical examination of the dogs with heart failure may reveal various clinical signs and symptoms (Fig. 13.). The clinical manifestations observed in cases of heart failure are related to forward failure, backward failure and arrhythmias. Clinical signs during early stage of heart failure are vague and nonspecific and seem related with decreased cardiac output. As heart failure progresses clinical signs, mentioned below, start appearing in different permutations and combinations.

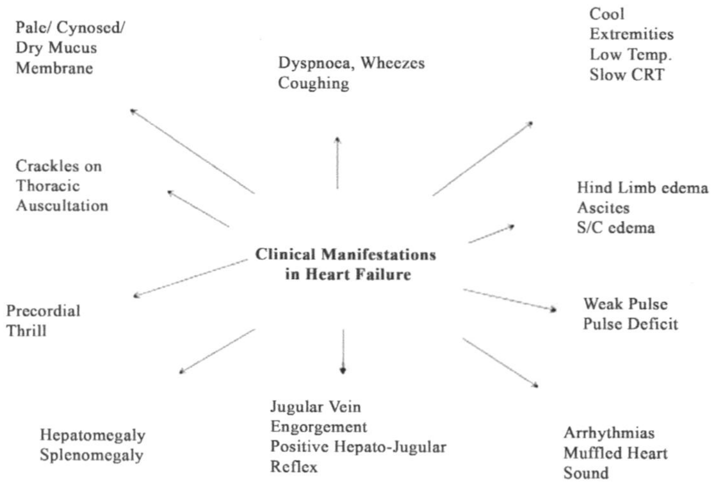

Fig. 13: Diagrammatic representation of clinical signs and symptoms in cases of heart failure.

- Pale, cyanosed (Fig.14 A), dry or tacky mucus membranes. Mucus membrane colour changes are related to low cardiac output or backward failure.
- Cool Extremities are due reduced cardiac output.
- Lowered rectal temperature
- Slow capillary refill time
- Engorgement of jugular vein (Fig. 14 B)
- Positive hepato-jugular reflux test-In a standing dog, pressure is placed in the region of cranial abdomen in a ventral to dorsal direction to check the presence of jugular distension or jugular pulsation. Positive hepato-jugular reflex is suggestive of increased pressure in the right atrium or right ventricle.
- Dyspnea, open mouth breathing (Fig. 14 C), wheezes
- Coughing
- Crackles on thorax auscultation
- Precordial thrills are detected on precordial palpation. The thrill are the vibrations on the chest wall due to fluid turbulence within the heart.
- Hepatomegaly (Fig.14 D) and/or splenomegaly

- Abdominal distension due to ascites (Fig.14 E). Ground glass appearance of abdomen radiographically (Fig. 14 F).
- Pleural Effusions (Fig. 14 G)
- Hind limb and /or subcutaneous edema (Fig .14 H)
- Weak femoral arterial pulse.
- Pulse deficit- Pulse rate lower than the heart rate is termed as pulse deficit. Many premature weak cardiac contractions do not lead to palpable pulse. Therefore, observation of pulse deficit creates a suspicion of occurrence of premature ectopic beats.
- Arrhythmia such as severe tachycardia, atrial fibrillations/flutter or ventricular tachycardia, ventricular fibrillations/flutters or severe heart blocks
- Muffled heart sound
- Decreased appetite / anorexia
- Cardiac cachexia (Fig.14 I)

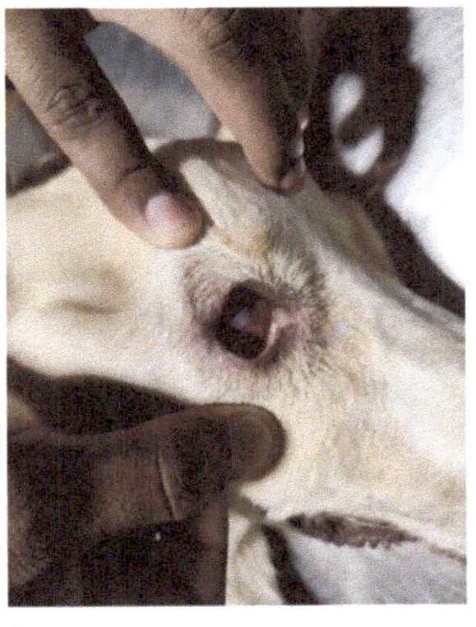

A

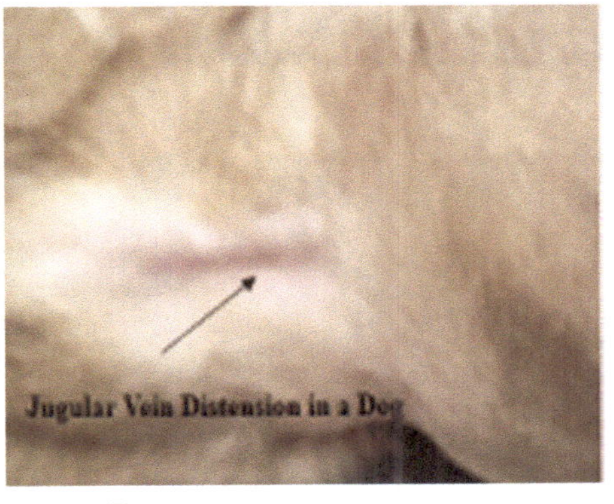

B

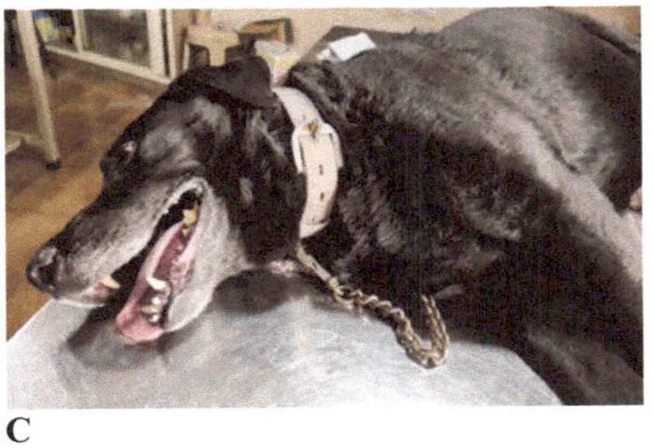

C

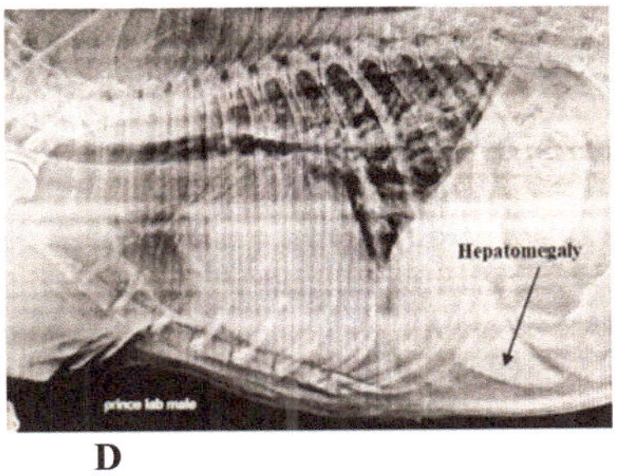

D

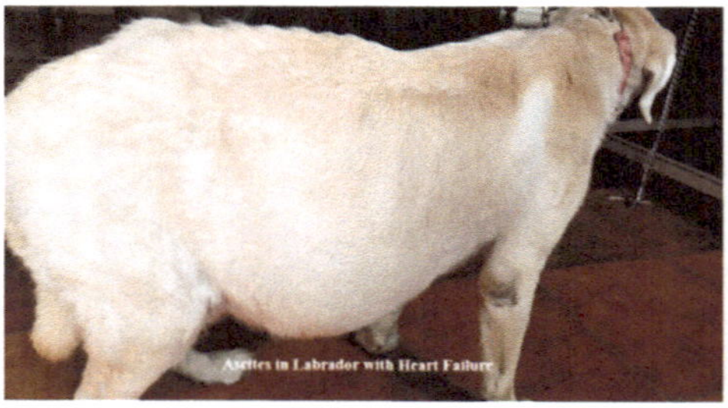

E

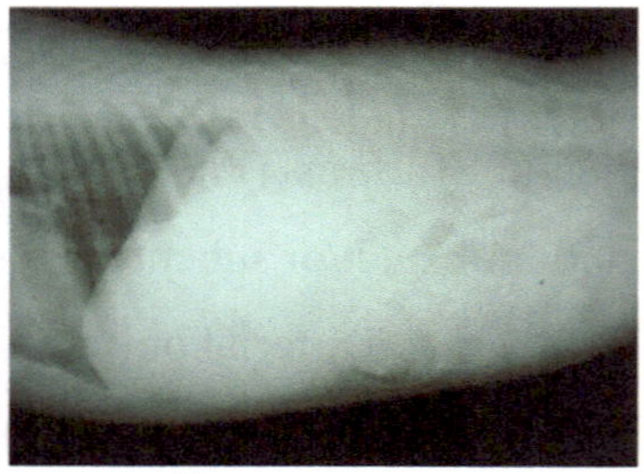

F

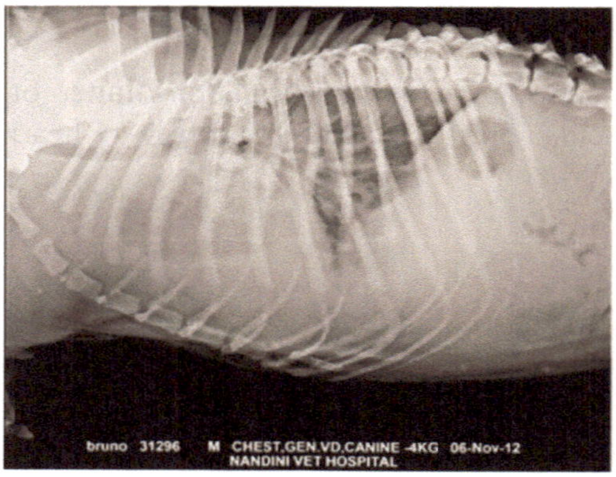

G

H

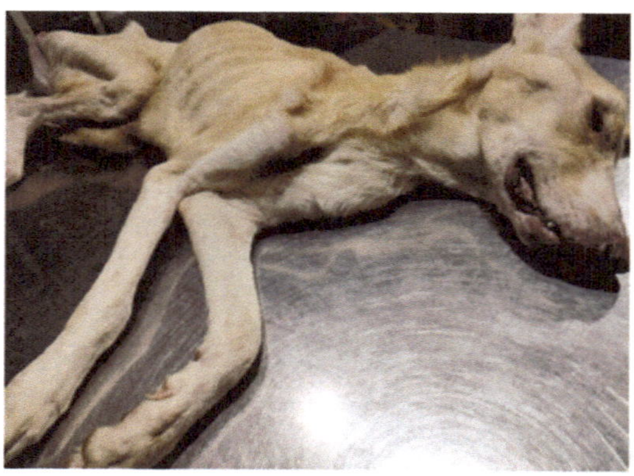

I

Fig. 14: Showing different clinical signs in dogs with heart failure. Cyanosed conjunctiva **(A)**, jugular distension **(B)**, open mouth breathing **(C)**, hepatomegaly **(D)**, ascites **(E)**, ground glass radiographic appearance of abdomen suggesting ascites **(F)**, pleural effusion **(G)**, limb edema **(H)** and cardiac cachexia **(I)**.

A small study conducted at this hospital has revealed that anorexia /reduced appetite, ascites, ataxia, cardiac gallop, cardiac murmurs, cold extremities, cardiac arrhythmias, coughing, dyspnea/orthopnea, hypothermia, jugular distension, changed colour of mucus membrane, prolonged capillary refill time, pronounced and variable apex beat, pulmonary crackle, weakness, weight loss ,weak femoral pulse, syncope and death accounted for 17.54, 14.0, 26.31, 61.40,21.05, 17.54, 52.63, 21.05,52.63, 15.78, 8.77, 22.80, 31.57, 28.07, 21.05, 63.15, 43.05, 29.83, 17.54 and 8.77 % cases of congestive heart failure in dogs due to heart diseases respectively. On 20 point scale of symptoms individual clinical score varied from 8 to 18 with a mean of 10 $\pm$0.61. (Varshney et al., 2011). Signs of congestive heart failure and low cardiac output may coexist in many cases.

Lung edema, coughing and dyspnea are the manifestations of left heart failure. The poor contractility of left heart in cases of left heart failure results into leakage of blood into left atrium through mitral valve eventually leading to congestion in the lungs. It is the congestion of the lungs that is in the core of respiratory signs in cases of heart failure. Fluid buildup in the abdomen (ascites) and /or in limbs (peripheral edema)are the result of right heart failure leading to poor venous return causing leakage of some blood into right atrium through tricuspid valve backing the blood into the circulation.

Weakness/lethargy/dullness/depression, exercise intolerance, syncope, cold extremities, hypothermia, dehydration/ increased capillary refill time, weak femoral pulse and pulse deficit are related to forward failure. While, persistent unproductive coughing, dyspnea, reluctance in lying down, distension of jugular vein , cyanotic or muddy mucus membranes, and fluid accumulation in different organs are related to backward failure.

Classification of Heart Failure

A. New York Heart Association (NYHA) has suggested classification of heart failure on the basis of heart function. It is known as Functional classification. It is classed in to four categories.

Class I. No symptom even on exercise

Class II. Fatigue, dyspnea or coughing on exercise

Class III. Signs mentioned in class II develops with minimum physical activity

Class IV. Signs such as dyspnea, coughing are seen even at rest and exaggerated with bare minimum physical activity.

B. Forrester's Classification- Heart failure has also been classified on the basis of cardiac output. This classification system is known as Forrester's classification.

Class I. Cardiac output and pulmonary venous pressure is normal

Class II. Cardiac output normal but there is pulmonary congestion

Class III. Decreased cardiac output, peripheral hypo perfusion but no pulmonary congestion

Class IV. Decreased cardiac output associated with congestion of the lungs.

C. Newer Classification System 2001-Consensus panel adapted the 2001 American College of Cardiology/American Heart Association Classification System for the treatment of heart disease and failure in human patients to the management of canine chronic valvular heart disease (Hunt *et al*., 2001). This system of classification is being used to compliment the functional classification system. This new system also has four basic stages of heart disease and heart failure.

Stage A. Patients having high risk of developing cardiac disease but at present no detectable structural abnormality of the heart .

Stage B. Structural heart abnormality (murmurs of mitral valve regurgitation) is detectable in patients but there is no clinical signs.

Stage B1. Patients are asymptomatic and there is no evidence of cardiac remodeling (radiographic or echocardiographic).

Stage B2. Patients are asymptomatic with radiographic and/or echocardiographic evidence of left heart enlargement.

Stage C. Patients are having signs of heart failure with evidence of structural heart disease.

Stage D. Patients with end stage disease with clinical signs of heart failure caused by chronic valvular heart disease (CVHD) refractory to standard therapy.

Diagnostic Approach in Heart Failure

Making a timely and exact diagnosis of heart failure in dogs and cats is not an easy task as no single diagnostic test is cent percent specific and sensitive. History provided by the owners and predominant clinical signs are not exclusive and are also seen in primary respiratory diseases. Therefore signs and symptoms of respiratory distress cannot be taken as pathognomonic of heart failure. Further, detection of heart failure in dogs in early stage may not be feasible because of absence of classical symptoms as symptoms of heart failure appear late warranting a compulsory systematic diagnostic approach. A thorough examination right from history; complete physical examination; radiographic examination; electrocardiographic examination; echocardiographic examination; to evaluation of cardiac biomarkers is necessary to arrive at the diagnosis of heart failure.

History

- History of sleep disruption, persisting nocturnal cough, dyspnea, marked weakness, increased panting in recumbency, exercise induced panting/ dyspnea and subcutaneous edema or ascites call for detailed cardiac evaluation to rule out/ rule in cardiac failure.

Clinical Examination

- Palpation of cardiac area may reveal precordial thrill due to fluid turbulence in the heart in cases of heart failure due to valvular insufficiency. Arrhythmias may also be detected.
- Auscultation of the heart is helpful in determining heart rate, detecting bradycardia or tachycardia, arrhythmias and murmurs or muffled heart sound.
- Marked weakness/lethargy, exercise intolerance, hypothermia, weak pulse, dehydration and sudden unconsciousness are related to low cardiac output due to forward failure.
- Signs of persistent unproductive coughing, respiratory distress, jugular vein distension effusions (peritoneal, pleural) and cyanotic/ muddy mucus membrane are related to backward failure.

Radiographic Examination

- Radiography is a valuable tool in providing an evidence of pulmonary venous congestion/ edema. Pulmonary venous distension occurs prior to cardiogenic pulmonary edema.
- Pulmonary veins (Fig.15) are evaluated on lateral thoracic radiographs. Generally pulmonary vein to cranial lung lobe is either equal or little larger than the accompanying pulmonary artery. When pulmonary veins to cranial lung lobe are thicker than 75% width of proximal one third of the fourth rib, it is said to be pulmonary venous distension.

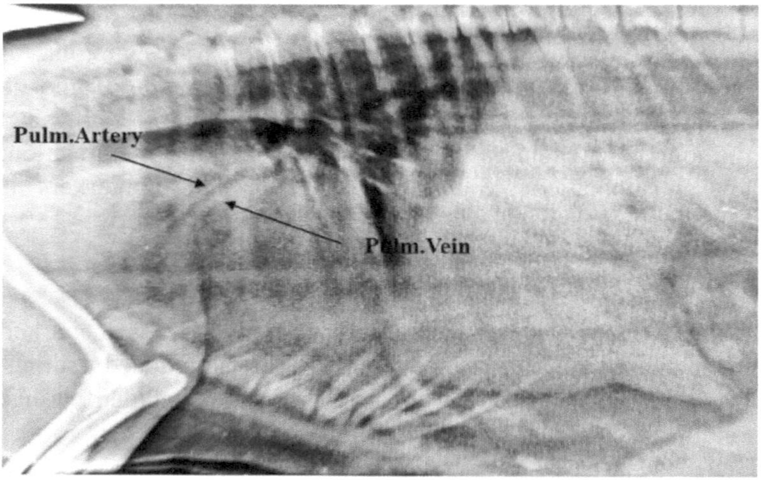

Fig. 15: Showing pulmonary artery and pulmonary vein in right lateral thorax radiograph of a dog.

- Detection of pulmonary edema in dorso-caudal hilar region (Fig.16) is suggestive of congestive heart failure in vast majority of cases. This edema appears first in central perihilar area .Then it progress outward dorsocaudally. Edema may also be seen in the anterior ventral area in the lateral radiographs. Cardiogenic pulmonary edema is due to excessive fluid accumulation in the interstitium and /or the pulmonary alveoli. In cats pulmonary edema occurs in a patchy, irregular pattern in the caudal lobes.

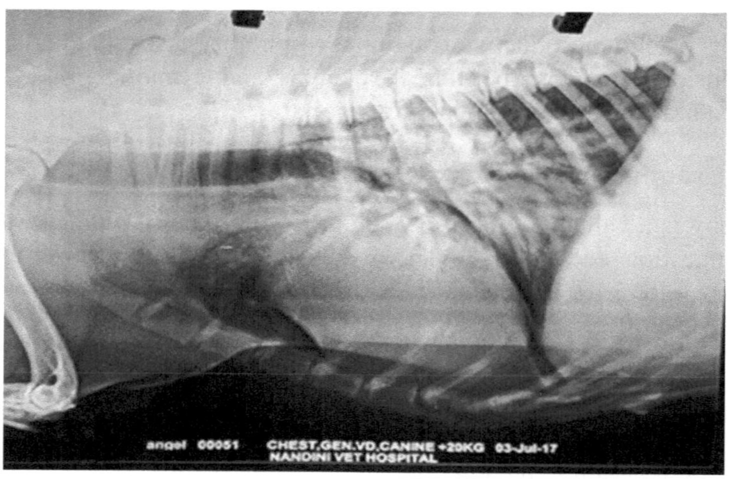

Fig.16 .Right lateral thoracic radiograph of a dog showing pulmonary edema in dorso-caudal hilar region suggesting congestive heart failure.

- Chamber enlargement, great vessel enlargement, heartworm disease, and effusions of pericardium or pleura (Fig.17) can also be detected on radiographic examination. In lateral radiographs normal heart occupies <2.5 intercostal space in deep chested dogs, <3 intercostal space in average size dogs, and < 3.5 intercostal space in brachiocephalic breeds, immature dogs or small breed dogs. Vertical distance from cardiac apex to carina is 2/3rd to 3/4th the vertical distance from cardiac apex to vertebral column. Reduction in right heart to right chest wall distance is suggestive of right ventricular enlargement. Similarly reduction in left heart to left chest wall distance is suggestive of left ventricular enlargement

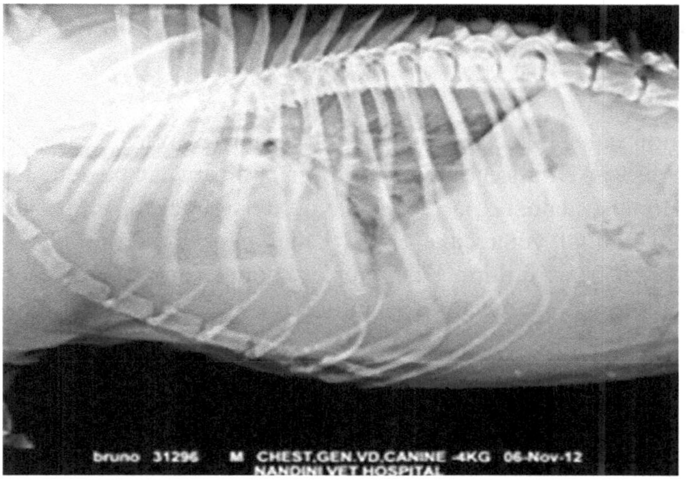

Fig. 17: Right lateral thoracic radiograph of a dog with heart failure showing pleural effusions.

- In lateral radiographs the heart appears more elongated and elliptical and ventricles occupy 2-2.5 intercostal space. In VD or DV views cat heart appears more oval and thin.

- Vertebral heart score (Fig.18) provides information about heart size. It is independent of respiration and position of the heart within the thorax. Vertebral heart score test has low sensitivity. Therefore normal vertebral heart score does not rule out heart enlargement. It also does not rule out heart diseases unrelated to heart size. Vertebral heart score in normal dogs and cats has been reported as 8.5to 10.7 and 6.7 to 8.1 respectively. Different breeds of dogs may have different VHS.

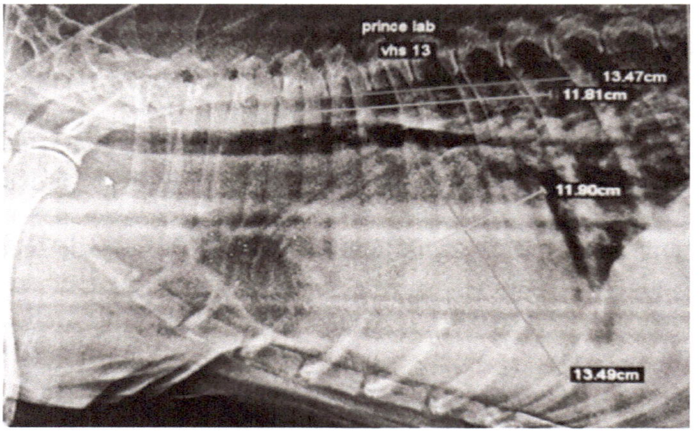

Fig. 18: Right lateral thoracic radiograph of a dog showing measurements of Vertebral Heart Score.

- Aortic distension can be seen in the region of aortic arch in cases of severe aortic stenosis. Dilation of main pulmonary artery can be detected if there is pulmonary stenosis, pulmonary hypertension, or dirofilariasis. Enlargement of caudal vena cava (Fig.19.) in lateral radiographs is suggestive of right heart failure. Diameter of caudal vena cava greater than the descending thoracic aorta is indicative of caudal vena cava enlargement.

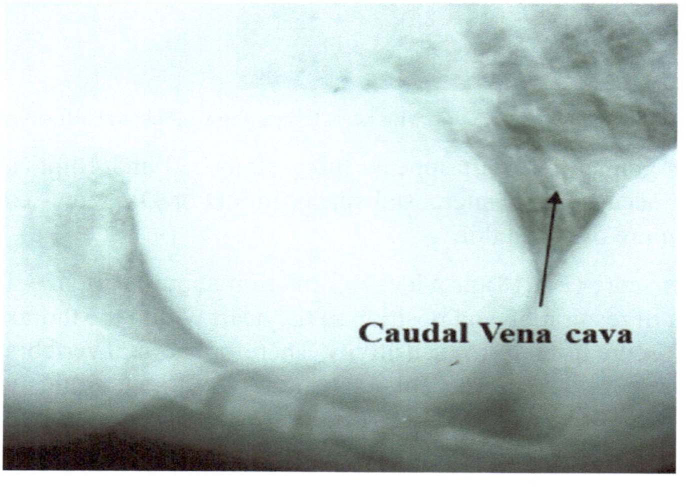

Fig. 19: Right lateral thoracic radiograph of a dog showing distension of caudal vena cava

- Globoid heart in lateral radiographs is generally suggestive of pericardial effusion. Pleural or peritoneal effusions (Fig.20) can also be detected in radiographs.

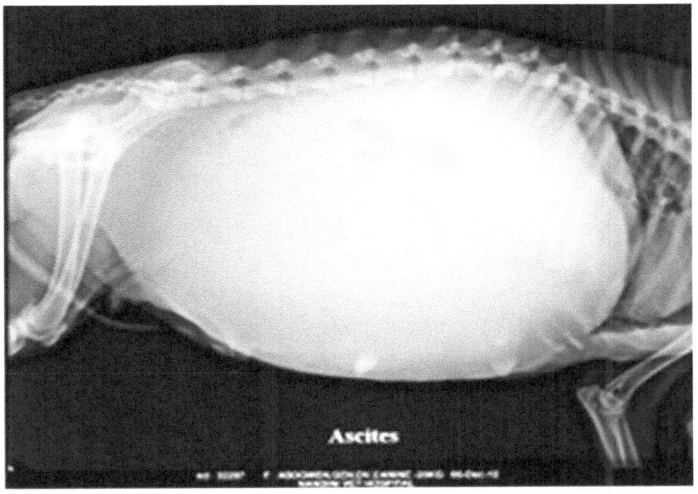

Fig. 20: Left lateral thoracic radiograph of a dog showing peritoneal effusions (ascites).

Electrocardiographic Examination

- Electrocardiography is another important diagnostic tool for ascertaining the presence or absence of the arrhythmias and conduction disturbances. Important arrhythmias (Fig. 21) are tachyarrhythmias, bradyarrhythmias, atrial fibrillation/flutter, atrial tachycardia, ventricular fibrillation/ flutter, ventricular tachycardia and severe 2nd / 3rd degree heart block. Electrocardiographic abnormalities may vary from sinus rhythm; sinus tachycardia/bradycardia; sinus arrest; atrial flutters/ fibrillation; atrial premature complex; atrial tachycardia; left atrial and/or left ventricular enlargement; ST elevation, ST depression to ventricular tachycardia (Varshney, 2020). Frequency and type of electrocardiographic abnormalities may vary in different breeds as ventricular arrhythmias are more common in Doberman and Boxers though atrial fibrillation are also seen in these breeds (Varshney *et al.*, 2011).

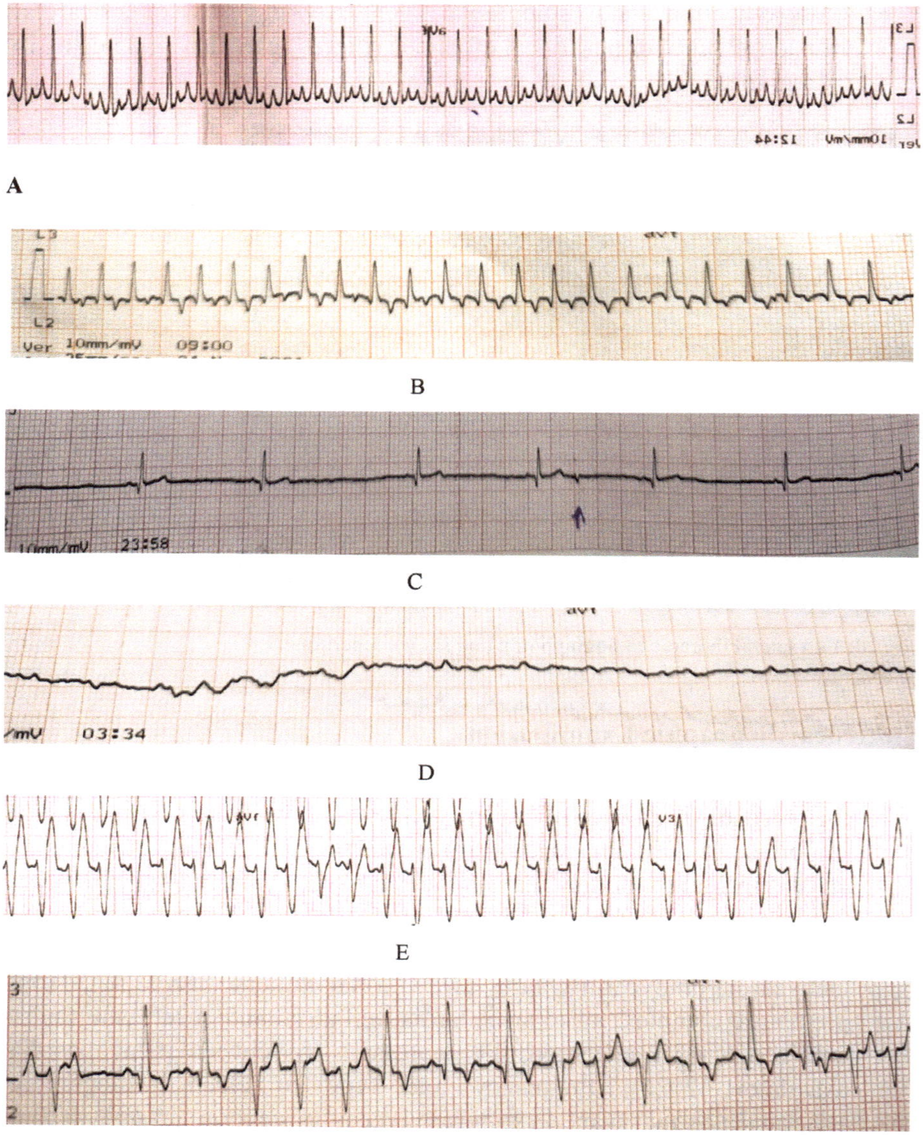

Fig. 21: Electrocardiograms of dogs (sensitivity 1, speed 25 mm/sec) showing various types of arrhythmias in dogs . A- Sinus Tachycardia (Heart rate 220 bpm with sinus rhythm) B-Atrial Fibrillations, C-Sinus Bradycardia (Heart rate 60 bpm with sinus rhythm), D-Ventricular fibrillations. E- Ventricular tachycardia, and F- ventricular premature complexes.

Echocardiographic Examination

- Echocardiography is the most convenient and appropriate means for detecting dilated or hypertrophic cardiomyopathy, septal defects, valvular defects, pericardial disease, cardiac tumors, bacterial endocarditis, systolic dysfunction (weakened cardiac muscles) or diastolic dysfunction (filling defect) leading to heart failure. Although echocardiography cannot definitely ascertain the presence of heart failure, it can provide evidence to support or refute the presence of heart failure. Blood flow abnormalities can be detected by Doppler echocardiography. Therefore Doppler echocardiography is most suitable means for detecting stenotic valvular disorders, valvular insufficiency septal defects, pulmonary artery hypertension, wall motion disorders etc. Saini and Varshney (2022) have described echocardiographic changes in dogs associated with different valvular insufficiencies and cardiomyopathies leading to heart failure.

Laboratory Examinations

- There is no definite blood test that can confirm the heart failure. Reduced cardiac output may be supported by increased level of serum creatinine, BUN and lactate. Decreased venous oxygen tension is supportive of reduced cardiac output. CPK, SGOT and LDH are of very little/no value in detecting heart disease in dogs and cats. Nevertheless isoenzymes are of some value in detecting myocardial necrosis.
- Recently cardiac troponin-I (cTn-I) has shown promise as a biomarker of cardiomyocyte injury. Increased level of cTn-I has been reported in primary heart diseases such as dilated cardiomyopathy, hypertrophic cardiomyopathy, sub aortic stenosis and chronic degenerative mitral valve disease; as well as in diseases causing secondary myocardial injury. Increased values of cTn-I (mean 5.19± 1.78 ng/ml, range 0.1 to 28.97 ng/ml) in dogs with primary left heart failure (Varshney *et al.*,2019) and in dogs with secondary myocardial insult in cases of canine monocytic ehrlichiosis (Varshney *et al.*,2015), babesiosis caused by B. gibsoni (Varshney 2017 a), heat stroke (Varshney, 2017 b) and in viper bites (Varshney and Monapara, 2019) have recently been reported.
- NT-pro BNP is another novel biomarker of cardiac insult and increased level has been reported in heart failure associated with cardiac diseases. NT-pro BNP have potential to differentiate respiratory signs associated with primary respiratory diseases or cardiac failure and can predict prognosis in cases of congestive heart failure. It can also be helpful in identifying occult cardiac disease prior to their onset. Cases having dilated cardiomyopathy seem to have higher NT-pro BNP level than those having degenerative mitral valve disease. Values of NT-pro BNP higher than 2700 pmol/l have been considered indicative of congestive

heart failure due to heart diseases. In dogs NT-pro BNP levels > 900 pmol/l may be suggestive of heart disease with or without heart failure.

- Higher levels of plasma norepinephrine , aldosterone and angiotensin II have been reported in cases of heart failure. Norepinephrine level seems to have prognostic value. Monitoring of digoxin level assists in titrating the dose of digoxin and management of arrhythmias developed on digoxin therapy. Estimation of electrolytes and acid base status can detect associated electrolyte and acid-base imbalances, if any. Thyroid disorders may also lead to heart failure in dogs and cats. Estimation of thyroid profile can be beneficial in addressing the primary cause in such cases. Blood culture yields desirable result in heart failure cases suspected with infective etiology. Snap test for microfilaria may assist in ruling out/ruling in cases of dirofilariasis as a cause of heart failure. Blood serum assays for taurine or L-carnitine may be of added advantage in case of heart failure due to dilated cardiomyopathy.

Assessing Probability of Heart Failure

Though , the diagnosis of the heart failure is not an easy task, the probability of heart failure can be assessed with a comprehensive systematic approach right from history to cardiac biomarkers. Diagrammatic representation of various diagnostic clues is presented in the Fig. 22.

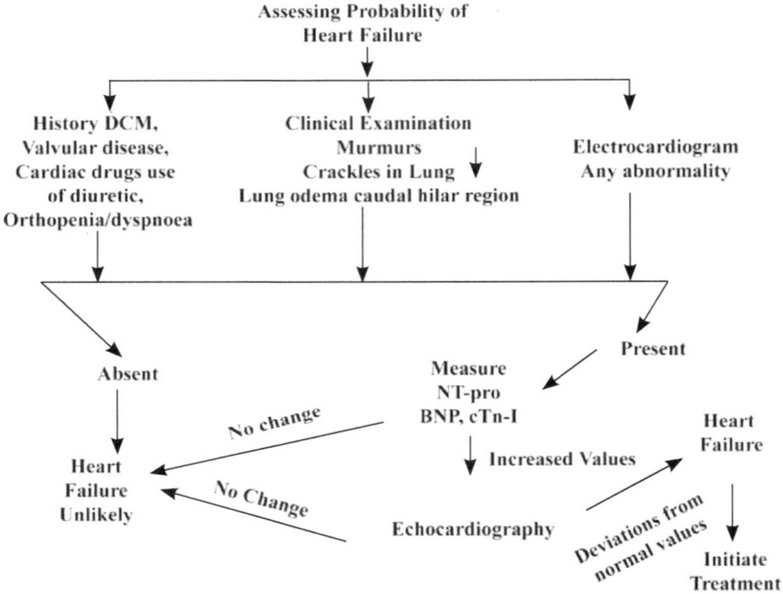

Fig. 22: Schematic representation of assessing probability of heart failure in dogs.

There is high probability of heart failure in dogs having history of heart diseases/heart failure, signs amenable to furosemide; presence of cardiac murmurs; crackle in lungs; edema in dorso-caudal hilar region of lungs; significant electrocardiographic abnormalities; increased levels of cTn-I and NT-pro-BNP; and echocardiographic evidence of cardiac structural and functional abnormalities.

History

- Previous history of heart failure/heart disease.
- History of furosemide amenable dyspnea/coughing in dogs

Clinical Signs

The following signs and symptoms may arise suspicion of heart failure.

- Dyspnea and loud cardiac murmurs in small breeds.
- Orthopnea and nocturnal coughing in dogs. Coughing in cats usually opposes the diagnosis of heart failure and is not a predominant sign of heart failure in cats.
- Positive hepatojugular reflex
- Jugular vein distension
- Gallop

Radiography

- Distended pulmonary veins (Left heart Failure). Pulmonary veins larger than pulmonary artery are considered as distended.
- Distended caudal vena cava (Right heart Failure).
- Perihilar interstitial to alveolar lung pattern is suggestive of left heart failure. Lung pattern is asymmetric in dogs with MVD and symmetric in dogs with DCM. Pulmonary pattern in cats is not similar to dogs. It is highly variable in cats with cardiogenic pulmonary edema.
- Pulmonary veins and cardiac silhouette may not be visible in dogs with severe heart failure due to severe diffuse interstitial to alveolar lung pattern or pleural effusions.
- Heart enlargement is not always consistent with heart failure.

Echocardiography

- Enlarged Atria
- Ratio of left atrium: aorta (>2.0)

Electrocardiography

- Left bundle branch block
- Atrial fibrillations/ Atrial tachycardia
- Ventricular fibrillations/ Ventricular tachycardia
- 3rd degree severe heart block

Cardiac biomarkers

- NT-pro BNP > 3000 pg/ml in dogs and > 1000 pg/ml in cats.
- cTn-I levels are higher in cases of heart failure in dogs and cats.

Prognosis of Heart failure

- Prognosis of heart failure due to hyperthyroidism, carnitine deficiency induced cardiomyopathy, taurine deficiency induced cardiomyopathy, and tachycardia induced heart failure is satisfactory if these conditions are reversed.
- Dogs with heart failure due to reasons other than described above usually die. Survival time in dogs with cardiomyopathy undergoing expert professional care varies from 6 months to 1 year.
- Small breeds of dogs with heart failure due to valvular insufficiency may survive from 1 to 2 years with rational therapy.
- Rupture of chordae tendinae in cases of severe chronic mitral valve disease may result in to episodes of severe fulminate pulmonary edema.
- Dobermans with heart failure due to cardiomyopathy in association with atrial fibrillation has poor prognosis.

Treatment of Heart Failure

- The main aims of treatment in heart failure is to enhance oxygenation, improve myocardial contractility, improve cardiac output, reduce pressure overload, reduce volume flow overload, reduce pulmonary congestion, overcoming restricted ventricular filling, reverting anorexia, managing arrhythmias, and managing oxidative stress (Fig.23).

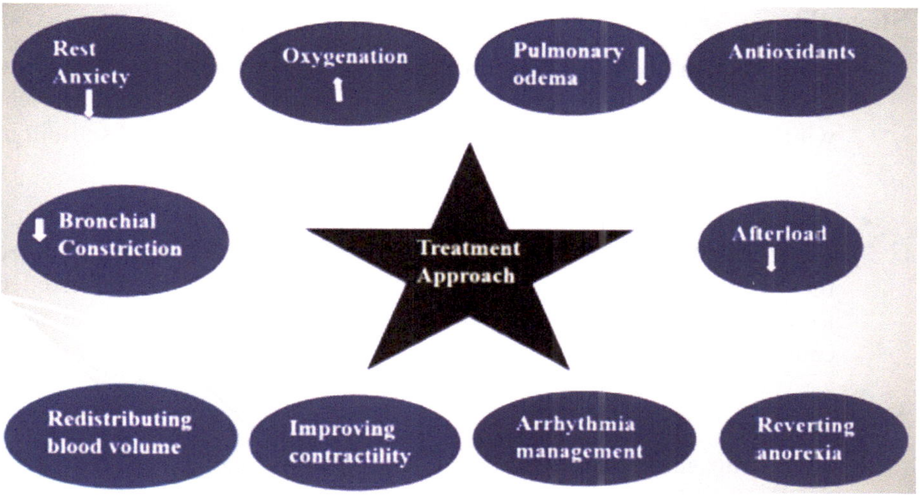

Fig. 23: Diagrammatic sketch of treatment approaches in heart failure.

- Further aims of treatment are different for short term (severely decompensated) versus long term (stable) heart failure patients and on the stage of the disease. As per guidelines by American Heart Association, heart failure in dog and cats has been categorized as stage A (animals at risk), stage B (structural abnormality is there but no signs of congestive heart failure), stage C (heart disease as well as signs of congestive heart failure both are present), and stage D (refractory congestive heart failure not amenable to standard therapy) as detailed under the heading classification of heart failure. Therapeutic interventions have been recommended as per the stage of heart failure.

Treatment as per Stage of Heart failure

Stage A

- No treatment needed.
- No dietary therapy needed.

Stage B1

- Patients with insignificant mitral regurgitation (no significant radiographic or echocardiographic enlarged LA and/or LV with normal LV systolic function, normal VHS, normal blood pressure and no significant biochemical changes) requires no drug and dietary therapy.
- Patients with Stage B1 needs to be re-evaluated every 6 to 12 months.

Stage B2

- Patients with significant mitral regurgitation with cardiac remodeling (enlarged LA, and/or LV, normal blood pressure) requires treatment with ACE-inhibitors, beta blockers, mild dietary restriction of sodium in small breeds.
- Treatment of dogs with dilated cardiomyopathy in asymptomatic phase may be started with ACE-inhibitors or beta blocker to minimize the progression of myocardial disease. Further studies are needed to substantiate this thinking. Beta blockers in dogs should be used cautiously.

Stage C

- Dogs with acute phase of heart failure require treatment with oxygen supplementation, diuretics, vasodilators, sedatives, and positive inotropes.
- Dogs with mild/moderate heart failure require treatment with furosemide, ACE-inhibitors, spironolactone, digoxin , pimobendan, diltiazem. (detail treatment of acute heart failure is given under the heading of Acute Heart Failure).

Stage D

- Doses of diuretics needs to be increased.
- Hydrochlorthiazide (0.5-1.0 mg/kg every 2^{n} or 3rd day) may be added.
- Amlodipine (0.1-0.4 mg/kg PO BID) can be added to reduce after load.
- Sildenefil (1-2 mg/kg PO BID/TID) is used in cases of advance congestive heart failure complicated by pulmonary hypertension (PH). Detail treatment of Stage D heart failure in given below under the heading of Refractory and Recurrent Heart Failure .

Treatment of Heart failure as per Severity

Acute Heart Failure

Acute heart failure is an emergency and requires emergency therapeutic management. Immediate concern in such cases is to increase oxygenation, to reduce fluid accumulation, to improve cardiac output, and to minimize stress.

1st Line Therapy in Acute Heart failure

a. *Oxygen Therapy*

- Supplemental oxygen can be given through flow by oxygen, nasal catheter, oxygen cage/ incubator. Details of oxygen therapy are given in chapter 3.
- Flow by oxygen should be started to the dogs with congestive heart failure during the physical examination as it is well tolerated and requires minimum restrain. As it is not very efficient in increasing the fraction of inspired oxygen, other methods (oxygen cages, oxygen hoods, nasal, nasopharyngeal and tracheal oxygen insufflation) need to be adopted for long term oxygen therapy.
- Humidified oxygen can be administered at the flow rate of 50 to 100 ml/ kg/minute. Non-humidified oxygen can be considered if a humidification source is not available. But non-humidified oxygen causes dryness and irritation of nasal mucosa making the dog uncomfortable and intolerant to nasal cannula.
- Oxygen saturation can be monitored by arterial blood sampling or pulse oxymetry. Pulse oxymetry reading can be taken on tongue, ear pinna, toe web, prepuce or vulva in cooperative dogs. Pulse oxymetry reading less than 94% requires administration of supplemental oxygen. Pulse oxymetry reading less than 80% in dogs receiving supplemental oxygen are the candidate for mechanical ventilation.

b. *Reducing Fluid Accumulations*

- The use of furosemide, a loop diuretic, is a corner stone of therapy of congestive heart failure. Fluid accumulation is reduced with the use of furosemide, but cardiac output is not improved. Dose of furosemide is to be tailored on individual basis. High doses of furosemide in cats and dogs with hypotension may deteriorate renal function and may cause electrolyte imbalance. On the other hand too low doses of furosemide do not reduce pulmonary edema and worsen the condition. Intravenous administration of furosemide is the better option in cases of severely decompensated heart failure. Results are better when administration of furosemide as a bolus is followed by constant rate infusion (CRI). In acute heart failure, the dose of furosemide varies from 2.0-4.0 mg/ kg intravenously or intramuscularly initially with hourly monitoring. If respiratory rate and respiratory efforts are not decreased, the dose of furosemide needs to be repeated as bolus or as constant rate infusion

(CRI) at the dose rate of 0.6-1.0 mg/kg/hr. intravenously. Respiratory rate, renal function and electrolyte are to be frequently monitored in dogs and cats undergoing furosemide therapy for heart failure. With the improvement in the condition of the patient the dose of the furosemide can be reduced to half. Once respiratory rate becomes normal, furosemide can be given orally.

- Torsemide is another diuretic of great utility in the management of acute severe refractory heart failure in dogs for its longer action, low susceptibility to diuretic resistance, and aldosterone antagonist property. Its dose is much lower than furosemide.
- Oral morphine, an opioid agonist, can be considered in dog with congestive heart failure @ 0.025 to0.05 mg/kg intravenously as it dilates splanchnic vasculature and increases venous capacitance permitting drainage of fluid from the pulmonary parenchyma making the respiration slow and deep. Morphine also decreases anxiety.
- Thoracocentesis is recommended in dogs and cats with severe respiratory distress due to large amount of pleural effusions.
- Intravenous fluid supplementation is contraindicated in dogs with acute fulminant congestive heart failure except in case of drugs being administered as CRI. In case of dire essentiality of fluid transfusion, low sodium fluid (Ringer's lactate solution having sodium 130 mEq/l), 5% dextrose in water, or half strength (0.45%) saline at lowest possible rate for drug administration may be considered.

c. *Improving cardiac output*

- For improving cardiac output in dogs with severe low cardiac output heart failure, dobutamine (1.0-10.0 μg/kg/min CRI), an adrenergic positive inotrope with beta-1 effect, is recommended. The dose may be increased with monitoring of blood pressure and electrocardiogram. Tachyarrhythmia (at higher dose rate), facial twitching and seizures are the potential side effects of dobutamine, therefore electrocardiographic monitoring should be continued.
- Pimobendan (0.25 mg/kg PO BID) is the drug of choice as soon as congestive heart failure is confirmed. Pimobendan, a phosphodiesterase-3 inhibitor, is an excellent inodilator that augment myocardial contractility by promoting calcium binding to cardiac troponin-C (cTn-C) without increasing oxygen demand of cardiomyocytes.

d. ***Minimizing stress***

- Strict cage rest should be provided to the dogs and cats with acute severe heart failure.
- Butorphanol (0.05 to 0.3 mg/kg IV/IM/SC) can be used for sedation purpose in dogs and cats with respiratory distress.
- Cats having pain due to aortic thromboembolism can be managed with buprenorphine.

2nd Line Therapy in Acute Heart Failure

- After 1st line therapy, monitoring of blood pressure is very important. Because of sympathetic stimulation blood pressure may increase in many dogs causing increase in the afterload.
- Hypertensive patients require vasodilators to reduce the afterload.
- Nitroglycerine (a vasodilator used transdermal) and nitroprusside (a potent venous and arterial dilator) are the additional drugs used in dogs and cats with severe dyspneic heart failure. Nitroprusside is not in common use nowadays. Effectiveness of topical nitroglycerine is also questionable.
- Hydralazine, an arteriodilator, can be used @0.5 to 3.0 mg/kg IV BID as bolus or @ 1.5-5.0 microgram/kg/min as CRI. Hydralazine needs to be used cautiously as reflex tachycardia and hypotension are its serious adverse effects requiring frequent monitoring.
- Hypotensive dogs (with DCM and /or DMVD) are to be treated with dopamine or dobutamine administered as CRI.

Follow up Treatment in Acute Heart failure

- Over next 24-48 hours as the patient stabilizes, intravenous/ CRI furosemide should be replaced with oral administration @ 2 mg/kg PO BID/TID for 3-4 days with evaluation of renal function.
- Cases should be evaluated (respiratory rate, thorax radiographs, electrocardiogram, renal and electrolyte profile) weekly for 2 to 3 weeks then monthly for 2-3 times then quarterly.
- Hypokalemic dogs should be supplemented with potassium.

Treatment of Chronic Heart Failure

- Main emphasis of treatment in chronic cases of heart failure revolves around dealing with deleterious effects of neurohormonal stimulation so that longevity can be extended and quality of life can be improved.
- Chronic cases of heart failure are managed with furosemide, pimobendan, ACE- inhibitor and spironolactone. These drugs are continued indefinitely.
- Furosemide is reduced gradually to a level of 1.0 mg/kg PO BID. The minimum dose of furosemide at which there is no respiratory distress is to be continued. Dogs on furosemide therapy should be evaluated for renal function as renal dysfunction may start in dogs on prolonged furosemide therapy.
- Pimobendan @ 0.25 mg/kg PO BID on empty stomach is to be continued.
- ACE- inhibitors in chronic heart failure with DCM and DMVD have extended longevity. Enalapril, benazepril, ramipril and quinapril are the ACE- inhibitors being used in dogs. Of these enalapril and benazepril are more common in use. Initially ACE-inhibitors are used once a day and their dose can be increased with the progression of the disease. Enalapril is excreted through kidneys. Therefore its use in renal compromised heart failure cases should be discouraged. In cases with renal- cardiac failure, benazepril is a better option as it is excreted 50% by kidneys and 50% by liver.
- Increased aldosterone level is detrimental to myocardium. Therefore, aldosterone receptor blockers such as spironolactone are of additional advantage. By giving spironolactone, the dose of furosemide can be reduced. Spironolactone is a weak diuretic. As a sole drug it is not effective in controlling heart failure. Spironolactone is used @ 1.0-2.0 mg/kg PO OD with food. Gynacomastia has been reported as a side effect of spironolactone in humans. There is no such report in dogs. Nevertheless potassium levels are required to be monitored in dogs as there may be hyperkalemia in dogs with spironolactone.
- Antiarrhythmic drugs are required to manage arrhythmias, if any. Ventricular arrhythmias can be managed with sotalol and mexiletine. Sotalol may exacerbate bradyarrhythmias. Gastrointestinal symptoms may be observed with mexiletine. Sustained atrial tachycardia and atrial fibrillation/flutters can be treated with a combination of digoxin and diltiazem. Beta blockers are frequently used to control heart rate in

human with heart failure. Because of their significant negative inotropic effect, their use in dogs with DCM having poor systolic function are not advisable. Details of antiarrhythmic therapy is given in chapter 9.

- Dietary modification is very important in the management of heart failure. It has been described in chapter 5.

Treatment of Refractory and Recurrent Heart Failure

- Dogs with refractory heart failure showing signs of pulmonary edema requires hospitalization.
- A higher dose of furosemide intravenously (two to three times) may be required to control pulmonary edema. Dose of furosemide exceeding 3-4 mg/kg TID may cause furosemide resistance. In such cases hydrochlorothiazide, another diuretic, needs to be added. Another option is to add torsemide, another loop diuretic. It is used @ one tenth (1/10th) of the total daily dose of furosemide. The total dose of torsemide should be divided into two and one half is given twice daily.

Newer Therapeutic Approaches in Heart Failure

Despite good understanding of heart failure and pharmacological developments in the management of chronic heart failure, the prognosis of dogs and cats with heart failure remains poor. Presently the treatment of heart failure is mainly directed against overstimulation of RAAS and is based on RAAS inhibitors, diuretics and inodilators extending the life span of the patients by few months. This fact has stimulated researches in detecting new pathways for testing new drugs to deals with heart failure beyond neurohumoral and hemodynamic modulatory therapeutic approaches. The development of new drugs for heart failure is focusing on the direct effect on cardiomyocytes, myocardial interstitium and coronary microcirculation.

- **RAAS Inhibitors**- RAAS inhibitors are well accepted drugs to reduce angiotensin-II, aldosterone, blood pressure or delaying end organ damage in congestive heart failure both in humans and dogs (Lefebre *et al.,* 2007). ACE-I (angiotensin converting enzyme-inhibitor) and ARBs (angiotensin receptor blockers) fall under RAAS inhibitors. ACE-Inhibitors prevent conversion of angiotensin I (AT-I) to angiotensin II (AT-II) and degradation of bradykinin. ARBs counteract AT-II at the level of AT-1 receptor. Enalapril, benazepril, imidapril and ramipril are the RAAS inhibiting drugs advocated in the management of congestive heart failure. Studies have shown that benazepril, highly potent selective inhibitor of ACE (angiotensin converting enzyme) , is able to reduce blood pressure of renal origin and is potent to significantly delay the time to onset of overt dilated cardiomyopathy in Doberman (O'Grady *et al.*,2009) .In renal compromised dogs with congestive heart failure, benazepril has advantage over enalapril that it is not solely excreted through kidneys . Therefore a better choice in cases with cardio-renal failure.

- **Phosphodiesterase III Inhibitor**- Pimobendan, a novel inodilator, has shown superior efficacy as compared to enalapril. Studies have shown that it can significantly delay the preclinical period in dogs with stage B2 myxomatous mitral valve disease.
- **Mineralocorticoid Receptor Antagonists (MRAs)** - Spironolactone is a mineralocorticoid receptor antagonist having diuretic property. It has been documented that spironolactone when used @ 2.0 mg/kg PO OD had significantly reduced the risk of cardiac morbidity and mortality in dogs with congested heart failure (Bernay *et al.*, 2010). Spironolactone with lower doses of furosemide can give better result. Spironolactone has been associated with gynecomastia in humans (Mosenkis and Townsend, 2004) because of its ability to bind to other steroid hormone receptors. Such information is not substantiated in dogs.
- **Torasemide** (Torsemide) is an another loop diuretic. It is more potent and has long lasting effect and is better than furosemide. Torasemide (0.24 mg/kg PO BID) has shown better results as compared to furosemide. It can significantly reduce myocardial fibrosis (Lopez *et al.*, 2004).
- **Old Targets and New drugs**
  (a) To overcome the side effects of spironolactone (gynecomastia in man), new more selective MRAs (eplerenone, finerenone) have been developed for the use in humans.
  (b) Renin Inhibitors (RIs) and Aldosterone Synthase inhibitors (ASIs), developed during last few years, have not shown encouraging results.

• New Targets and New Drugs

Angiotensin Receptor Neprilisyn Inhibitors (ARNI) have been designed to block newly identified angiotensin receptor neprilisyn target (ARNI). Sacubitril/valsartan is the first drug in this class and has been recommended for the treatment of heart failure with reduced ejection fraction (Yancy *et al.*,2013). This drug has been found superior to enalapril in humans. Preliminary studies (Mochel *et al.*,2014, 2019) using sacubitril / valsartan (675mg/day) have shown substantial increase in cGMP and significant decrease in plasma aldosterone level in heart failure .Trial conducted by Newhard *et al.* (2018) with sacubitril/valsartan (20 .0 mg/kg PO BID) in dogs with myxomatous mitral valve disease (MMVD) has shown significant reduction in urinary aldosterone to creatinine ratio and increased survival time.

- **Development of New Drugs in Humans with Heart Failure**

New pharmacotherapeutic modalities (Omecamtive Mecarbil,

Empaglifozin,CD-NP,Urocortin-2,) are being investigated in human heart failure with reduced ejection fraction (HfrEF).

(a) *Omecamtive Mecarbil (OM)*- This drug differ from other inotropes as it does not increase intracellular calcium (Ca2+), no concomitant increase in oxygen consumption but improves myocardial contractility (Selim *et al.*,2017). It is a myosin binding peptide and can assist in increasing myocardial contractility.

(b) *Tolvaptan*- It, a vasopressin antagonist, may be the drug of choice for heart failure as it reduces pre and after load by decreasing systolic blood pressure and blood volume.

(c) *Empaglifozin (EG)*- This is an antidiabetic drug acts as an osmotic diuretic to reduce blood pressure (Tikkanen *et al.,* 2015). The drug has shown a lower rate of cardiovascular death and heart failure hospitalization in type -2 diabetic patients (Fitchett *et al.,* 2016).

(d) Metformin – It is another antidiabetic drug that has potential to reduce the risk of heart failure and also improves the clinical course of heart failure in humans (Charytan *et al.*,2019). Exact mechanism of metformin is at present not known completely. Generally many of the characteristics common to diabetes and heart failure such as insulin resistance, endothelial dysfunction, inflammation and oxidative stress have shown improvement with metformin.

(e) *Thiazolidinediones (*TZDs)- It is another antidiabetic drug that reduces CVD risk but may cause fluid gain and fluid retention.

(f) *SERCA2 Activator*- It is a specialized Ca2+ pump responsible for calcium re uptake in sarcoplasmic reticulum and is being investigated in human with heart failure (Jessup *et al*., 2011).

(g) *CD-NP*- It is a synthetic Natriuretic peptide (NP) causing a vasodilation with a minimum effect on blood pressure and has also shown inhibitory effect on myocardial fibrosis in end stage heart failure in humans (Ichiki *et.al,* 2014).It also decreases ALD production in healthy humans (Lee *et al*., 2009).

(h) *Urocortin-2-* It is a member of corticotrophin releasing factor (CRF) family and has high affinity to CRF receptor. It has shown promising results in improving myocardial function in animal models of heart failure; and also in clinical cases of acute heart failure in humans (Chan *et al.*,2013).

(i) *Serelaxin and Ularitide-* These drugs are next generation vasodilators having potential to reduce peripheral vascular resistance and to increase cardiac index.

• Identifying New Targets and New Drugs for Heart Failure in Humans

Concerted efforts are being made to improve the outcome of heart failure in humans through identifying new targets in target areas and developing more effective therapies. The development of new drugs for heart failure in humans is being focused on recognizing additional/new targets such as cardiomyocytes, myocardial interstitium, microcirculation and inflammation. The subject of addressing new targets in heart failure in humans is being reviewed during recent years .Some new developments are briefly enumerated below.

(a) In the area of microcirculation, EC (cardiac microvascular endothelial cell) metabolism and EC nutrient transport are the area being targeted and translation drugs are being investigated.

(b) In the area of myocardial interstitium, myocardial interstitial fibrosis and chymase are being targeted for development of new drugs. Sacubutril/Valsartan, Empagliflozin and Fulacimstat are the drugs being investigated.

(c) Cardiomyocyte calcium handling, cardiomyocyte nitroxyl donors, cardiomyocyte mitochondria, signaling pathway and myofibrillar pathway are the area under cardiomyocyte that are being investigated for the development of new cardiac drugs. The new targets in this area are SERCA2a, partial adenosine A1 receptor, cytochrome C, soluble guanylate cyclase (sGC), arginine vasopressin signaling, and myosin istaroxime,

(d) Inflammation is another area in which the main target is interleukin 1 receptor. Canakinumab and Anakinra are the new drugs being investigated.

Gene Therapy in Heart Failure

- During recent years gene therapy of heart failure is gaining momentum. It is amongst the most promising approaches of reversing the fundamental abnormalities in failing cardiomyocytes.
- Gene therapy permits manipulation of molecular targets that are presently refractory to other therapeutic interventions.
- Gene therapy offers the possibilities to interfere with the molecular pathways governing the pathogenesis of the heart failure.
- Gene therapy is a procedure resulting in the transfer of the gene into an individual's cells to treat a disease. The introduction of genetic material into the cells not only compensate for abnormal genes but also influences pathways involved in the development of disease process.
- The goal of the gene transfer is to express a functional gene when endogenous gene is dysfunctional.
- Heart failure being a complex syndrome characterized by multiple defects at cellular level, gene therapy aims at myocardial protein levels to improve myocardial function.
- The history of cardiac gene therapy dates back to 1990 (Lin *et al.*, 1990). When rat cardiomyocytes were transfected, in vivo, by injecting plasmid DNA containing the beta-galactisodase gene directly in left ventricular wall.
- Gene therapy was initially envisioned as a treatment strategy for inherited monogenic disorders. It has now broad therapeutic potential including treatment of acquired polygenic diseases such as peripheral vascular disease, ischemic heart disease, arrhythmias, and congestive heart failure (Tilemann *et al.*, 2012).
- With the better understanding of the molecular mechanism associated with heart failure and vectors with cardiotropic properties, gene therapy is now being considered a viable adjunct to mechanical and pharmacological therapies for heart failure. It is a novel stratergy for targeting abnormalities in cardiac cells that adversely affect the myocardial function.
- New vectors (adeno- associated virus) for gene delivery can result in sustained transgene expression.
- Reduction in SERCA2a activity (cardiac form of sarco endo plasmic reticulum calcium2+ ATPase) causes abnormal handling of calcium by

the failing heart leading to systolic and diastolic dysfunction. The up regulation of calcium by percutaneous administration of gene therapy in cardiac disease has shown promising results.

- Gene transfer therapy in cases of heart failure can be performed by either direct introduction into the heart or presentation to the myocardium via circulation (Tilemann *et al*., 2012).
- Gene therapy targets in heart failure in humans have been identified as SERCA2a, phospholamban, G-protein coupled kinase-2 receptors, protein phosphatase 1, adenylate cyclase 6 (AC-6) (Vinge *et al*.,2008) and stromal cell derived factor 1 (SDF-1).
- Decreased expression and activity of SERCA2a in myocardium has been identified in animal models of heart failure as well as human patients with heart failure (Mercadier *et al*., 1990, Gupta *et al*., 1990). Restoration of SERCA2a expression and activity by gene transfer has shown improvement in cardiac function, reversion of left ventricular remodeling and survival rate in humans (Niwano *et al*., 2008, Kawase *et al*.,2008).
- In heart failure, expression and activity of AC-6 decreases. Adenoviral-mediated AC6 gene transfer had shown to reduce left ventricular remodeling and increase rates of survival in both small and large animal models of heart failure (Lai *et al*., 2004; Tang *et al*., 2004; Rebolledo *et al*., 2006).
- SDF-1 is other strategy that has been investigated in human heart failure. It plays important role in tissue repair.SDF-1 gene transfer therapy has shown potential efficacy in humans with heart failure (Penn *et al*., 2013).
- Sleeper et al.(2004) treated six dogs with mucopolysaccharidosis VII(MPS VII), a lysosomal storage disease caused by deficient beta-glucuronidase activity resulting in defective catabolism of glucosaminoglycans (GAGs) leading to cardiomyopathy owing to accumulation of GAGs, intravenously with a retro virus (RV)- expressing canine GUSB (beta-glucuronidase) and found highly encouraging results.

Chronopharmacotherapy

- Chronopharmacotherapy, comparatively a new idea, relates with the timing of drug administration for optimum efficacy with minimum adverse effect. An understanding of circadian rhythms can have impact on the therapeutic management of RAAS related diseases. Chronopharmacotherapy has been used in the treatment of rheumatoid arthritis (Staessen *et al.*, 1992), lung cancer (Mazzoccoli *et al.*, 2012) and cardiovascular diseases in humans. Investigations have revealed that ACE- inhibitors in humans had shown greater reduction in blood pressure when given in evening as compared to morning dose schedule (Hermida andAyala, 2009). Study conducted by Sole and Martino (2009) has shown that heart and vessel growth occurred more actively during normal sleep time. Experimental studies in mice have also confirmed improved cardiovascular functioning, and reduced adverse cardiac remodeling when captopril, an ACE inhibitor, was given to mice at sleeping hours (Martino et al.,2011). These studies have confirmed more effective results of ACE inhibitors at night time in humans with hypertension. Very meagre information about systemic dynamics of renin cascade is available in dogs. Studies conducted on chronobiology of RAAS in relation to blood pressure and sodium potassium handling in dogs (Mochel *et al.*, 2013a, 2014) have revealed that sodium intake influenced the mesor and phasic secretion of renin (with greater sodium intake, the amplitude and mesor of renin angiotensin-RA is smaller). It has also been observed that the time of sodium intake also had synchronizing effect on the acrophase of RA and blood pressure oscillations in dog (Itoh *et al.*,1996). It appears logical that the optimized efficacy of cardiac drugs would be with bedtime (night) dosing in dogs also.

- High sodium intake in humans is thought to play a significant role in the genesis of blood pressure, cardiovascular and renal diseases. Based on this thinking sodium restriction has been a common recommendation in dogs with congestive heart failure. Substantial evidence on association of higher sodium intake and increased risk of hypertension in dog is lacking (Anderson *et al.*, 1986; Greco *et al.*, 1994). Some studies have revealed that the mesor and amplitude value of RA oscillation are greater in dogs fed on low sodium regime (Mochel *et al.*, 2014 b). It seems that normal diet (not sodium restricted diet) will be more beneficial in dogs with congestive heart failure.

References

Adin, D.B.,Taylor,A.W. , Hill,R.C., Scott,K.C., and Martin ,F.G. (2003). Intermittent bolus injection versus continuous infusion of furosemide in normal adult greyhound dogs. J.Vet.Intern. Med. 17:632-636.

Anderson, D.E., Gomez-Sanchez, C. and Dietz, J.R. (1986).Suppression of plasma renin and aldosterone in stress-salt hypertension in dogs. Am. J. Physiol. 251(1 Pt 2):R181-186.

Arkadievich,O.D.(2020).Metabolic markers of myocardial insulin resistance in dogs with heart failure. Open Vet.J. 10:363-370.

Ashrafian,H.,Frenneaux,M.P. and Opie,L.H. (2007). Metabolic mechanisms in heart failure. Circulation116:434-448. https://doi.org/10.1161/Circulation AHA.107.702795.

Bernay, F., Bland, J.M., Häggström, J., Baduel, L., Combes, B., Lopez, A. and Kaltsatos ,V. (2010).Efficacy of spironolactone on survival in dogs with naturally occurring mitral regurgitation caused by myxomatous mitral valve disease. J. Vet. Intern. Med. 24:331-341.

Byrne, J.A., Grieve,D.J., Cave, A.C.,and Shah, A.M. (2003).Oxidative stress and heart failure. Archives des Maladies du Coeur et des Vaisseaux 96: 214-221. (Pub Med).

Calvert, C.A., Pickus, C.W. and Jacobs, G.J. (1997) . Signalment, Survival and Prognostic \ in Doberman Pinschers with end stage cardiomyopathy. J. Vet. Intern. Med. 11 : 323 -331.

Chan, W.Y., Frampton, C.M., Crozier, I.G., Troughton, R.W. and Richards, A.M. (2013). Urocortin-2 infusion in acute decompensated heart failure: findings from the UNICORN study (urocortin-2 in the treatment of acute heart failure as an adjunct over conventional 834 therapy). JACC Heart Fail. 1:433-41.

Charytan, D.M., Solomon, S.D., Ivanovich, P., Remuzzi, G., Cooper, M.E., McGill, J.B., Parving, H., Parfrey, P., Singh, A.K., Burdman, E.A., Levey, A.S., Eckardt, K., McMurray, J.J.V. , Weinrauch, L.A., Liu,J., Claggett , B., Lewis, E.F. and Pfeffer, M.A. (2019). Metformin use and cardiovascular events in patients with type 2 diabetes and chronic kidneydisease. Diabetes Obes. Metab. 21:1199-1208. doi: 10.1111/dom.13642. Epub 2019 Mar 4.

Chobanian, A.V., Bakris, G.L., Black, H.R., Cushman, W.C., Green, L.A., Izzo, J.L. Jr., Jones, D.W., Materson, B.J., Oparil, S., Wright, J.T. Jr. and Roccella, E.J. (2003). Joint National Committee on Prevention, Detection, Evaluation, and Treatment of High Blood Pressure. National Heart, Lung, and Blood Institute; National High Blood Pressure Education Program Coordinating Committee. Seventh report of the Joint National Committee on Prevention, Detection, Evaluation, and Treatment of High Blood Pressure. Hypertension. 42:1206-52"

Cremer,S.E.,Singletary,G.E.,Olsen,L.H.,Wallace,K.,Haggstrom,J.,Ljungvall,I.,Hoglund,K., Reynold, C.A.,Pizzinat, N. and Oyama, M.A. (2014). Serotonin concentrations in platelets, plasma, mitral valve leaflet, and left ventricular myocardial tissue in dogs with myxomatous mitral valve disease. J. Vet. Intern. Med. 28:1534–1540. doi: 10.1111/ jvim.12420.

Cunningham, S.M., Rush, J.E. and Freeman, L.M. (2012).Systemic inflammation and endothelial dysfunction in dogs with congestive heart failure. J. Vet. Intern. Med. 26:547–557. [PubMed]

Dekker, R.L., Moser, D.K., Tovar, E.G., Chung, M.L., Heo,S., Wu, J.R., Dunbar, S.B., Pressler, S.J. and Lennie, T.A. (2014). Depressive symptoms and inflammatory biomarkers in patients with heart failure. Eur.J .Cardiovasc.Nurs. 13:444–450. https://doi.org/10.1038/ jid.2014.371

Domanjko,P.A.,Lukman,T.,Verk,B. and Nemec,S.A.(2018). Systemic inflammation in dogs with advanced stage heart failure.Acta.Vet.Scand.60:16-18. https://doi.org/10.1186/ s13028-018-0372-x.

Dooley, R., Harvey, B.J. and Thomas, W. (2012). Non-genomic actions of aldosterone: From receptors and signals to membrane targets. Mol. Cell. Endocrinol. 350: 223- 234.

Ferasin,L.,Fritz,M., Ferasin,H., Becquart,P.,Corbet,S.,Gouilh, M.A.,Legros,V. and Leroy,E.M. (2021).Infection with SARS-CoV-2 variant B.1.1.7 detected in a group of dogs and cats with suspected myocarditis. Vet.Rec. e944 :1-9. https://doi.org/10.1002/vetr.944.

Fonfara,S.,Hetzel,U.,Tew,S.R.,Cripps, P., Dukes-McEwan, J. and Clegg, P.D.(2013).Myocardial cytokine expression in dogs with systemic and naturally occurring cardiac diseases. Am.J.Vet.Res.74: 408-416. https://doi.org/10.2460/ajvr.74.3.408.

Fonfara, S., Hetzel, U., Oyama, M.A. and Kipar A.(2014). The potential role of myocardial serotonin receptor 2B expression in canine dilated cardiomyopathy. Vet. J. 199:406–412. [PubMed] [Google Scholar]

Greco, D.S., Lees, G.E., Dzendze,l G. and Carter, A.B.(1994). Effects of dietary sodium intake on blood pressure measurements in partially nephrectomized dogs. Am. J. Vet .Res. 55:160-165.

Güder, G., Bauersachs, J., Frantz, S., Weismann, D., Allolio, B., Ertl, G., Angermann, C.E. and Störk, S.(2007). Complementary and incremental mortality risk prediction by cortisol and aldosterone in chronic heart failure. Circulation. 115:1754-1761.

Gupta, R. C., Shimoyama, H., Tanimura, M., Nair, R., andLesch, M. (1990). SR Ca2+- ATPase activity and expression in ventricular myocardium of dogs with heart failure. , J.Am J Physiol 273, 12-18. PubMed

Hermida, R.C. and Ayala,D.E.(2009).Chronotherapy with the angiotensin-converting enzyme inhibitor ramipril in essential hypertension: improved blood pressure control with bedtime dosing. Hypertension. 54:40-46.

Hunt,S.A.,Baker,D.W.,Chin, M.H., Cinquegrani, M.P., Feldman, A.M., Francis, G.S.,Ganiats, T.G., Goldstein, S., Gregoratos, G., Jessup,M.L. , Noble,R.J., Packer.M., Silver,M.A., Stevenson, L.W., Gibbons, R.J., Antman, E.M., Alpert, J.S., Faxon, D.P., Fuster,V., Jackobs, A.K., Hiratzka, L.F., Russell, R.O. and Smith Jr., S.C. (2001). ACC/AHA guidelines for the evaluation and management of chronic heart failure in the adult: Executive summary. A report of the American College of Cardiology/American Heart Association Task Force on Practice Guidelines (committee to revise the 1995guidelines for the evaluation and management of heart failure). J. Am. Coll. Cardiol.38:2101–2113.

Ichiki, T., Schirger, J.A., Huntley, B.K., Brozovich, F.V., Maleszewski. J.J., Sandberg, S.M., Sangaralingham, S.J., Park, S.J. and Burnett, J.C. Jr.(2014). Cardiac fibrosis in end-stage human heart failure and the cardiac natriuretic peptide guanylylcyclase system: regulation and therapeutic implications. J. Mol. Cell Cardiol. 75:199-205.

Ide, T., Tsutsui,H., Kinugawa, S. , Utsumi,H.,Kang, D., Hattori, N.,Uchida,K., Arimura,Ki., Egashira,K., and Takeshita,A.(1999). Mitochondrial electron transport complex I is a potential source of oxygen free radicals in the failing myocardium. Circulation Res. 85:357-363.(PubMed) (Google Scholar).

Itoh, K., Kawasaki, T. and Cugini, P. (1996).Effects of timing of salt intake to 24-hour blood pressure and its circadian rhythm. Ann. N Y Acad. Sci. 783: 324–325.

Janeiro, M. H., Ramirez, M. J., Milagro, F. I., Martinez, J. A. and Solas, M.(2018). Implication of trimethylamine N-oxide (TMAO) in disease: potential biomarker or new therapeutic target. Nutrients 10: 1398

Janus, I., Kandefer-Gola, M., Ciaputa, R., Noszczyk-Nowak, A., Paslawska, U., Tursi, M. and Nowak, M.(2017).Cardiomyocyte marker expression in dogs with left atrial enlargement due to dilated cardiomyopathy or myxomatous mitral valve disease. Folia Histochem. Cytobiol. 55:52- 61.

Jessup, M., Greenberg, B., Mancini, D., Cappola,T., Pauly, D.F., Jaski, B., Yaroshinsky, A., Zsebo, K.M., Dittrich, H. and Hajjar, R.J. (2011). Calcium Upregulation by Percutaneous Administration of Gene Therapy in Cardiac Disease (CUPID) Investigators. Calcium Upregulation by Percutaneous Administration of Gene Therapy in Cardiac Disease (CUPID): a phase 2 trial of intracoronary gene therapy of sarcoplasmic reticulum Ca2+- ATPase in patients with advanced heart failure. Circulation. 124:304- 313.

Karlin, E. T., Rush, J. E. and Freeman, L. M. A.(2019).Pilot study investigating circulating trimethylamine N-oxide and its precursors in dogs with degenerative mitral valve disease with or without congestive heart failure. J. Vet. Intern. Med .33: 46–53.

Kawase, Y., Ly, H. Q., Prunier, F., Lebeche, D. and Shi, Y. (2008).Reversal of cardiac dysfunction after long-term expression of SERCA2a by gene transfer in a pre-clinical model of heart failure.J. Am.Coll.Cardiol. 51 :1112-19.

Lai, N. C., Roth, D. M. ,Gao, M. H., Tang, T. and Dalton, N. (2004). Intracoronary adenovirus encoding adenylyl cyclase VI increases left ventricular function in heart failure. J. Circulation 110: 330-336.

Lee, C.Y., Chen, H.H., Lisy,O., Swan, S., Cannon, C., Lieu, H.D. and Burnett, J.C. Jr. (2009). Pharmacodynamics of a novel designer natriuretic peptide, CD-NP, in a first-in human clinical trial in healthy subjects. J.Clin.Pharmacol. 49:668-673.

Lefebvre, H.P., Brown, S.A., Chetboul, V., King, J.N., Pouchelon, J.L. and Toutain, P.L. (2007). Angiotensin-converting enzyme inhibitors in veterinary medicine. Curr. Pharm. Des. 13:1347-1361.

Levant,P., Kocaturk, M., Akgun, E., Saril,A.,Cevik,O.,Baykal,A.T.,Tanaka,R.,Ceron,J.J. andYilmaz,Z .(2020).Platelet proteome changes in dogs with congestive heart failure. BMC Vet.Res.16:466. https://doi.org/10.1186/s12917-020-02692-x

Lin, H., Parmacek, M.S., Morle, G., Bolling,S. and Leiden,J.M.(1990). Expression of recombinant genes in myocardium in vivo after direct injection of DNA. Circulation 82: 2217-2221. Doi:10.1161/01CIR 82.6.2217.

Lopaschuk,G.D.,Ussher,J.R. ,Folmes,C.D.S.,Jaswal,J.S. and Stanley,W.C. (2010).Myocardial fatty acid metabolism in health and diseases. Physiological Reviews 90. https://doi. org./10.1152/physrev.00015.2009.

Lopez,B., Querejeta,R., Gonzalez,A.,Sanchez,E., Larman,M. and Diez,J. (2004). Effect of loop diuretics on myocardial fibrosis and collagen type I turnover in chronic heart failure. J. Am.Coll.Cardiol. 43: 2028-2035.

Mak, S. and Newton,G.E. (2001). The oxidative stress hypothesis of congestive heart failure: Radical thought. Chest 120:2035-2040. https://doi.org/10.1378/chest.120.6.2035.

Mallery, K.F., Freeman, L.k., Harpster, N.K. and Rush, J.E. (1999). Factors contributing to the euthanasia decision in dogs with congestive heart failure. Am. J. Vet. Med. Assoc. 214:1201-1204.

Martino, T.A., Tata ,N., Simpson, J.A, ,Vanderlaan, R., Dawood, F., Kabir, M.G., Khaper, N., Cifelli, C., Podobed, P., Liu, P.P., Husain, M., Heximer, S.,Backx,P.H. and Sole, M.J.(2011). The primary benefits of angiotensin-converting enzyme inhibition on cardiac remodeling occur during sleep time in murine pressure overload hypertrophy. J. Am. Coll.Cardiol. 57:2020-2028.

Mazzoccoli, G., Sothern, R.B., Parrella, P., Muscarella, L.A., Fazio, V.M., Giuliani, F., Polyakova, V. and Kvetnoy, I.M .(2012). Comparison of circadian characteristics for cytotoxic lymphocyte subset in non-small cell lung cancer patients versus controls. Clin.Exp. Med. 12: 181–194.

McMichael, M.A.,Freeman,L.M., Selhub,J., Rozansk,E.A., Brown,D.J., Nadeau,M.R. and Rush,J.E. (2000). Plasma homocysteine, B vitamins and amino acid concentrations in cats with cardiomyopathy and arterial thromboembolism.J.Vet.Intern.Med. 14:507-512.

Mercadier, J, J,,Lompré, A, M,, Duc, P., Boheler, K. R., andFraysse, J, B. (1990).Altered sarcoplasmic reticulum Ca2+-ATPase gene expression in the human ventricle during end-stage heart failure. , (1990)J.JClin Invest 85, 305-309.PubMed

Mochel, J.P., Burkey, B., Garcia, R., Peyrou, M., Giraudel, J., Renard, D. and Danhof, M. (2014). First-in class angiotensin receptor neprilysin inhibitor LCZ696 modulates the dynamics of the renin cascade and natriuretic peptides system with significant reduction of aldosterone exposure. J .Am. Coll.Cardiol. 63(12_S). doi: 10.1016/S0735-1097(14)60806-8

Mochel, J.P., Fink, M., Peyrou, M., Desevaux, C., Deurinck,M., Giraudel, J.M.and Danhof, M. (2013). Chronobiology of the renin-angiotensin-aldosterone system in dogs: relation to blood pressure and renal physiology. Chronobiol. Int. 30 :1144-59.

Mochel, J.P., Fink ,M., Bon, C., Peyrou, M., Bieth, B., Desevaux, C., Deurinck, M., Giraudel, J.M. and Danhof, M.(2014). Influence of feeding schedules on the chronobiology of renin activity, urinary electrolytes and blood pressure in dogs. Chronobiol.Int. 31:715-730.

Mochel, J.P., Teng, C.H., Peyrou, M., Giraudel, J., Danhof, M. and Rigel, D.F. (2019).Sacubitril/ valsartan (LCZ696) significantly reduces aldosterone and increasescGMP circulating levels in a canine model of RAAS activation. Eur. J. Pharm. Sci. 128:103-111.

Moesgaard, S.G., Soresen, T.M., Sterup, A., Tarnow,I., Kristensen, A.T., Jensen, A.L., and Olsen, A.H. (2009). Changes in platelet function in dachshunds with early stage ofmyxomatous mitral valve disease. Res.Vet.Sci.86:320-324.

Mosenkis, A. and Townsend, R.R. (2004).Gynecomastia and antihypertensive therapy. J.Clin. Hypertens (Greenwich).6:469-470.

Newhard, D.K., Jung, S., Winter, R.L. and Duran, S.H.(2018). A prospective, randomized, double blind, placebo-controlled pilot study of sacubitril/valsartan (Entresto) in dogs with cardiomegaly secondary to myxomatous mitral valve disease. J. Vet. Intern. Med. 32 :1555-1563.

Niwano, K., Arai, M., Koitabashi, N., Watanabe, A. and Ikeda, Y. (2008).Lentiviral vector-mediated SERCA2 gene transfer protects against heart failure and left ventricular remodeling after myocardial infarction in rats. J. Mol.Ther. 16:1026-1032.PubMed

O'Grady, M.R. and Horne, R. (1995) . Outcome of 103 asymptomatic Doberman Pinschers: Incidence of dilated cardiomyopathy in a longitudinal study. J. Vet. Intern. Med.9 : 199.

O'Grady, M.R., O'Sullivan, M.L., Minors, S.L. and Horne, R.(2009). Efficacy of benazepril hydrochloride to delay the progression of occult dilated cardiomyopathy in Doberman Pinschers. J. Vet. Intern. Med. 23:977-983.

Penn, M .S., Mendelsohn, F .O., Schaer, G. L., Sherman, W. and Farr, M. (2013) An open-label dose escalation study to evaluate the safety of administration of nonviral stromal cell-derived factor-1 plasmid to treat symptomatic ischemic heart failure.J Circ. Res. 112 :816-825.

Petrič, A.D., Lukman, T., Verk, B. and Svete, A.N.(2018). Systemic inflammation in dogs with advanced-stage heart failure. Acta Vet. Scand. 60:1. [PMC free article] [PubMed] [Google Scholar]

Prihirunkit,K.,Sastravaha,A.,Lekcharoensuk,C. and Chanloinapha, P.(2014). Hemoststic markers in congestive heart failure dogs with mitral valve disease.J.Vet.Med.2014:589873. doi:10.1155/2014/589873. PMCID:PMC4590876.

Rebolledo, B., Lai, N. C., Gao, M. H., Takahashi, T. and Roth, D. M. (2006) Adenylylcyclase gene transfer increases function of the failing heart. J. Hum. Gene. Ther.17,1043–1048. Pub Med

Reimann,M.J.,Ljungvall,I.,Hillstrom,A.,Moller,J.E.,Hangman,R.,Falk.T.,Hoglund,K., Haggstrom,J. and Olsen,L.H. (2016). Increased serum C-reactive protein concentrations in the dogs with congestive heart failure due to myxomatous mitral valve disease. Vet.J.209:113-118.https://doi.org/10-1016/j.tvjl.2015.12.006

Romito,G., Bertaglia, T., Bertaglia, L., Decaro,N.,Uva, A.,Runga, G., Morenc, A.,Vincifori,G., Dondi,F.,Diana,A. and Cipone, M. (2021).Myocardial injury complicated by systolic dysfunctionin a Covid-19 positive dog.Animals 11, 3506. https://doi.org/10.3390/ani11123506

Rubio, C.P., Saril,A.,Kocaturk,M.,Tanaka,R.,Koch, J.,ceron,J.j. and Yilmez, Z. (2020). Changes in inflammatory and oxidative stress biomarkers in dogs with different stages of heart failure. BMCVet.Res.16: 43 . https://doi.org/10.1186/s12917-020-02650-7.

Schulthesiss,H.,Fairweather,D.,Caforio,A.L.P.,Escher,F.,Hersheberger,R.E.,Lipshultz,S.E., Liu,P.P.,Matsumori,A., Mazzanti, A., McMurray,J. and Priori,S.G. (2019). Dilated cardiomyopathy.Nat. Rev. Dis.Primers5:32.https://doi.org/10-1038/s41572-019-0084-1.

Saini, N. and Varshney, J.P (2022).Cardiac Ultrasound . In: Ultrasound in Veterinary Medicine: Fundamentals and Application. Varshney,J.P. (ed.), NIPA, a division of NIPA Genx Electronic Resources and Soluttions P. Ltd., New delhi-110034.P179- 213..

Seligmann,H.,Halkin,H.,Rauchfleisch,S.,Kaufmann,N.,Tal,R., Motro,M., Vered,Z.V.I. and Ezra,D. (1991). Thiamine deficiency in patients with congestive heart failure receiving long term furosemide therapy: a pilot study. Am.J.Med. 91:151-155.

Selim, A., Zolty, R. and Chatzizisis,Y.S. (2017).The evolution of heart failure with reduced ejection fraction pharmacotherapy: What do we have and where are we going? Pharmacol.Ther. 178:67-82.

Seo,J., Matthewman,L.,Xia,D., Wilshaw,J.,Yu-MeiChang and Connolly,D.J.(2020). The gut microbiome in dogs with congestive heart failure: a pilot study. Sci.Rep. 10:13777. https://doi.org/10.1038/s41598-020-70826-0.

Sisson, D.D. and Thomas W.P. (1995) . Myocardial Diseases. In : Text book of Veterinary Internal Medicine. Ettinger, S.J. and Feldman, E.C. (Edn), 4thEdn. W.B. Sanders Company, Philadelphia. pp. 995 – 1032.

Sleeper, M.M., Fornasari, B., Ellinwood, N.M., Weil, M.A., Melniczek, J., O'Malley, T.M., Sammarco, C.D.,Xu,L.,Ponder,K.P. and Haskins, M.E. (2004).Gene therapy ameliorates cardiovascular disease in dogs with mucopolysaccharidosisVII. Circulation 110: 815-820. doi: 10.1161/01.CIR.0000138747.82487.4B. Epub 20.

Sole, M.J. and Martino, T.A. (2009). Diurnal physiology: core principles with application to the pathogenesis, diagnosis, prevention, and treatment of myocardial hypertrophy and failure. J .Appl. Physiol. 107:1318-1327.

Staessen, J., Guo, C., De Cort, P., Fagard, R., Lijnen, P., Thijs, L., Van Hoof, R. and Amery, A.(1992). Mean and range of the ambulatory pressure in normotensive subjects. Chin. Med. J. (Engl) 105:328-33.

Tikkanen, I., Narko, K., Zeller, C., Green, A., Salsali, A., Broedl, U.C. and Woerle, H.J.(2015). EMPA REG BP Investigators. Empagliflozin reduces blood pressure in patients with type 2 diabetes and hypertension. Diabetes Care. 38:420-428.

Tang, T., Gao, M. H., Roth, D. M. ,Guo, T. and Hammond, H. K. (2004) .Adenylyl cyclase type VI corrects cardiac sarcoplasmic reticulum calcium uptake defects in cardiomyopathy. , J.Am. J. Physiol. Heart Circ. Physiol. 287: 1906-1912. PubMed

Tangmahakul,N.,Makoom,P. and Surachetpong, S.D.(2021). Assessment of platelets and plasma serotonin in canine pulmonary hypertension secondary to degenerative mitral valve disease. Front Vet.Sci.15. https://doi.org/10.3389/fvets.2021.695492.

Tilemann, L., Ishikawa, K., Weber, T.and Hajjar, R.J.(2012). Gene therapy for heart failure. Circ Res. 110 :777–793.

Varshney, J.P. (2017a). Cardiac evaluation in anaemic dogs with clinical babesiosis caused by *Babesia gibsoni*. Indian J.Vet.med. 37:14-16.

Varshney,J.P.(2017 b). Evaluation of myocardial damage in dogs with heat stroke. J.Anim.Res. 7: 979-981.

Varshney,J.P. (2018). Right heart failure in an adult Pomernian suffering from fatal dirofilariasis. Blue Cross Book37:95-97.

Varshney, J.P.(2020). Electrocardiography in Veterinary Medicine. Springer, Singapore

Varshney,J.P.,Deshmukh, V.V. and Chaudhary, P.S. (2011). Congestive heart failure in dogs- A prospective study. Intas Polivet 12 : 261-267.

Varshney, J.P., Deshmukh, V.V. and Chaudhary, P.S. (2015). Evaluation of myocardial injury in acute canine monocytic ehrlichiosis. Intas Polivet. 16: 340-344.

Varshney, J.P., Huma,Z.I and Sharma, N. (2019). Evaluation of cardiac troponin-I and vitamin D3 in dogs with left heart failure and its management with pimobendan. Vet.Pract.20 :74-77.

Varshney,J.P. and Monapara, H.D. (2019). Evaluation of cardiac toxicity in dogs bitten by viper snake. Vet.Prac. 20: 191-194.

Vatnikov,Yu.,Rudenko,A.,Rudenko,P.,Kulikov, Ev., Karamyan,A.,Lutsay,V., Medvedev, I., Byakhova,V., Krotova,E. and Molvhanova, M.(2019). Immune-inflammatory concept of the pathogenesis of chronic heart failure in dogs with dilated cardiomyopathy. Vet. World 12:1491-1498.

Vinge, L.,Raake,P. and Koch, W.(2008). Gene therapy in heart failure. Circ.Res. 102: 1458-1470.

Yancy, C.W., Jessup, M., Bozkurt, B., Butler, J., Casey, D.E. Jr., Colvin, M.M., Drazner. M.H., Filippatos, G., Fonarow, G.C., Givertz, M.M., Hollenberg, S.M., Lindenfeld, J., Masoudi, F.A., McBride, P.E., Peterson, P.N., Stevenson, L.W. and Westlake, C (2016). 2016 ACC/AHA/HFSA Focused Update on New Pharmacological Therapy for Heart Failure: An Update of the 2013 ACCF/AHA Guideline for the Management of Heart Failure: A Report of the American College of Cardiology/American Heart Association Task Force on Clinical Practice Guidelines and the Heart Failure Society of America. J.Am.Coll.Cardiol. 68 : 1476-88.

Zavalishina, S.Y,,Kutafina, N.V., Vatnikov, Y.A., Makurina, O.N., Kulikov, E.V., Rystsova, E.O., Gurina, R.R. and Sotnikova, E.D.(2016). Platelet-activity dependence on the age of rats with experimental dyslipidemia. Biol. Med. (Aligarh) 8:6. [Google Scholar]

Zeisel,S.H. and Warrier,M. (2017). Trimethylamine N-oxide, the microbiome, and heart and kidney disease. Annu.Rev.Nutr.37:157-181.

Zois,N.E.,Moesgaard,S.G.,Kjelgaard-Hansen,M.,Rasmussen,C.E.,Falk,T.,Fossing,C.,Haggstrom,J.,Pederson,H.D. and Olsen, L.H. (2012). Ciculating cytokine concentrations in dogs with different degrees of myxomatous mitral valve disease. Vet.J.192:106-111. https://doi.org/10-1016/j.tvjl.2011.05.009.

3

Oxygen Therapy

Oxygen therapy has been in vogue for more than century and it is key component in cardiorespiratory care both in human and veterinary medicine. It is recommended in all hypoxemic humans, dogs and cats (peripheral oxygen saturation < 90% to 94% or PaO2 < 70 mm Hg) presented with acute heart failure or respiratory diseases. But oxygen therapy in normoxemic subjects has remained controversial as some studies have shown detrimental effect (increased level of reactive oxygen species, oxidative stress, reduction in coronary blood flow and myocardial oxygen consumption) of oxygen therapy in subjects with normal oxygen level. Therefore, oxygen saturation level should be monitored before instituting oxygen therapy. Oxygen therapy is a lifesaving short term therapy used to stabilize the patient. This therapy in itself is not a curative therapy but it is used only to stabilize the patients.

Oxygen Therapy

- Oxygen is very vital to life. Cell death, organ failure and death may occur in any animal depending on the severity of oxygen depletion in the body. Therefore, extra oxygen needs to be supplemented when there is oxygen hunger/ oxygen demand in body.
- The aim of oxygen therapy is to supplement the air the animals breath in with extra oxygen.
- Oxygen therapy is a lifesaving therapy in animals (canine , feline, rabbits, calves, pet birds) not only with acute heart failure but also with severe respiratory distress owing to other causes. Hypoxia in animals with severe anemia is not responsive to oxygen therapy.

Indications

- Acute heart failure
- Dyspnea/increased respiration/severe panting
- Shortness of breath/ open mouth breathing
- Cyanotic gums

- Prolonged recumbancy
- Severe hemorrhages
- Pulse oximetery reading < 93 -94 %
- Arterial partial pressure of oxygen <70 mm Hg (Mazzaferro, 2015)

Clues for Oxygen Therapy

- Tachypnea
- Shortness of breath
- Orthopnea
- Cyanosis
- Respiratory noises.
- Hypoxia. There is a need to differentiate hypoxia /respiratory distress of cardiac origin from other causes as animals with severe ascites, pain, metabolic acidosis or on certain drugs (opioid or corticosteroids) may breathe heavily or with great effort.

There are four types of hypoxia viz. hypoxic hypoxia (seen in hypoventilation, pulmonary diseases, patent ductus arteriosus, or pulmonary fibrosis), anaemic hypoxia (seen in immune mediated hemolytic anemia, hemorrhages), circulatory hypoxia (seen in hypovolemic shock, septic shock, anaphylaxis, cardiac tamponade, systolic failure, massive pulmonary thromboembolism), and histotoxic hypoxia (mitochondrial dysfunction, cyanide poisoning). Therefore a thorough clinical examination and detailed analysis of history is rewarding. Hypoxia of cardiac and respiratory origin responds well to oxygen therapy. Respiratory distress due to traumatic brain injury also responds well to oxygen therapy.

Type of Oxygen Therapy

- Short Term- It does not require humidification.
- Long Term- It requires humidified oxygen. Flow rate can be adjusted to 50-100 ml/kg/minute.

Measurements of Oxygen Saturation

- Pulmonary edema causes severe impairment of pulmonary diffusion leading to hypoxia in dogs with congestive heart failure.

- Oxygen saturation can be measured by non-invasive pulse oximetry or measuring arterial blood oxygen.
- Oxygen measuring should be done without causing undue stress to the ailing patient.
- When oxygen measuring is stressful, in such cases administer oxygen and oxygenation status can be gauged by reduction in respiratory rate and change in respiratory efforts and resolution of pulmonary crackles on chest auscultation.
- In clinically stable dogs with heart failure, oxygen saturation can be measured with pulse oxymetry or arterial blood sampling.
- For arterial blood gas analysis, blood sample can be obtained from dorsal pedal or femoral arteries.
- If a dog undergoing oxygen supplementation shows PaO_2<60 mm Hg and $PaCo_2$> 60 mm Hg, then dog should be shifted to mechanical ventilation.
- Pulse oximetry readings can be attempted on the tongue, ear pinna, toe web, prepuce, or vulva in cooperative patients. It is most convenient stress free and non-invasive method of monitoring oxygen delivery to the tissues. It calculates the percent of oxygen saturation of hemoglobin in arterial blood using spectrophotometry (Ford and Mozzaferro, 2006).

Methods of Oxygen Delivery

- Methods of oxygen delivery falls in two categories as non-invasive and invasive. Non-invasive methods of oxygen therapy consists of flow by oxygen, face mask, oxygen hood/oxygen collar, and oxygen incubator / oxygen cage. Invasive methods of oxygen therapy consists of nasal oxygen prongs, nasal oxygen catheter, and transtracheal oxygenation.

Non-Invasive methods of Oxygen Therapy

- Flow by oxygen administration- It is the simplest way of oxygen administration. In this technique, the end of hose of oxygen is placed adjacent to or within 2 cm of patient's nose or mouth. There is no need of humidifier in this method. Pets generally tolerate this mode of oxygen administration. This method provides 40% maximum fraction of inspired oxygen. Flow by oxygen is well tolerated and minimum restrain is required. But it is relatively inefficient method of providing an increase in the fraction of inspired oxygen. A flow rate of 2 to 3 liter/

minute is maintained which provides a FiO2 of 25 to 40 % (Loukopoulos and Reynolds, 1997).

- Oxygen administration by mask- Face mask (Fig.24 A) of various sizes are available. Mask provides tight fit and a higher concentration of oxygen can be administered from a oxygen cylinder (Fig. 24 B) as compared of flow by. It is to be ensured that there is a gap between patient's nares and oxygen delivery point and nares should not be squashed to the end of the mask. Oxygen therapy using face mask is easy and convenient and can be used in dogs (Fig. 25 A), cats and calves (Fig. 25 B). Oxygen cylinder (Fig. 24 B) or centralized oxygen source (Fig. 25 C) can be used as a source of oxygen. Many dogs are uncomfortable with mask and do not cooperate. The nervous or uncooperative dogs are not ideal candidate for this mode of oxygen administration. Oxygen saturation depends on the fitness of the mask. Carbon dioxide may build up when mask is very securely fit resulting into re-breathing of carbon dioxide. Such case requires periodic aeration. A flow rate of 2-5 liter /min can provide a FiO_2 of up to 50-60% (Boyle, 2012).

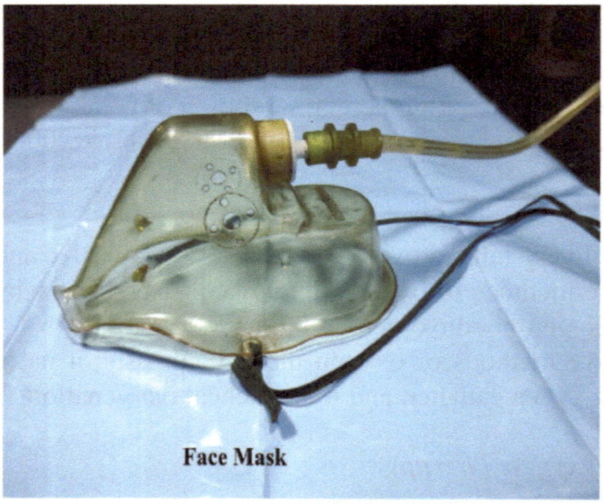

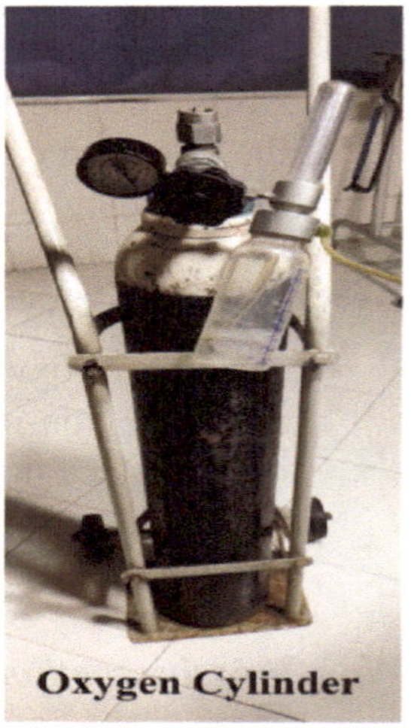

A B

Fig. 24: Showing Face Mask (A) and Oxygen Cylinder (B)

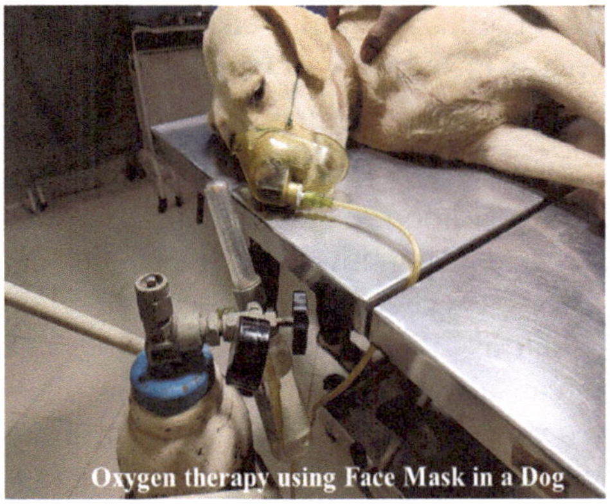

A

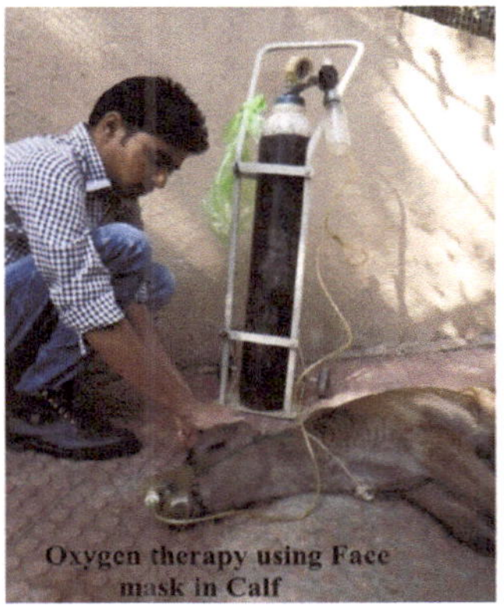

B

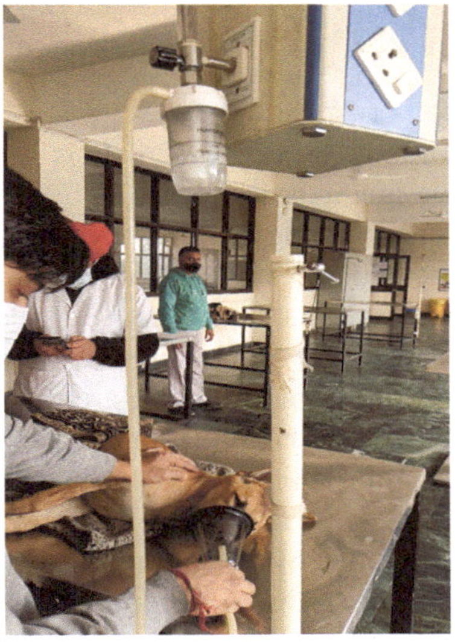

C

Fig. 25: Showing oxygen therapy in dog and calf using face mask. The source of oxygen is oxygen cylinder in Fig .A and B; and centralized oxygen supply in Fig. C. Figure C is with the courtesy of Dr. Neetu Saini, Associate Professor, Department of Veterinary Medicine, Guru Angad Deo Veterinary and Animal Science University, Ludhiana.

- Oxygen administration through oxygen cage /incubators – It is an excellent method of supplying oxygen to small size animals in stress free environment. Medical pediatric incubator or commercially available portable oxygen cages can be used. Depending on the size of the cage or incubator, a flow rate of 2 to 10 L/min. should be maintained. While administrating oxygen through cage/incubator, temperature and humidity inside the cage should be regularly monitored as closed cage heats up and there is rapid increase in humidity. Oxygen cages with internal controls are better option. This method provides 60% maximum fraction of inspired oxygen.

- Oxygen administration through oxygen hood /oxygen collar- It is very effective method of oxygen delivery using low flow rates. Oxygen hoods are suitable for large size dogs .Oxygen hoods are commercially available. With little resources such as Elizabethan collar, white tape and plastic wrap oxygen hoods can also be created in the hospitals. Oxygen through hoods is tolerated well .But increased condensation and iatrogenic hyperthermia may develop. Temperature and humidity is to be controlled. Flow rate of 0.5 to 1.0 L/min. should be maintained. This method provides FiO_2 of 40-50 %. To prevent drying of eyes, ophthalmic ointments are to be used.

Invasive Method of Oxygen Therapy

- Nasal Oxygen-It is a better way of administrating oxygen. It is the best inexpensive method among invasive methods of oxygen delivery. Catheters are available in various sizes and length. The oxygen is delivered by nasal cannula or French red rubber catheter (No. 5 or 8) placed in nasal cavity and secured with tape/sutures. One or two cannula or catheter may be used. This method provides 60% maximum fraction of inspired oxygen. Nasal or nasopharyngeal oxygen cannulae are well-tolerated for long-term oxygen supplementation. The tube is marked and tip is lubricated with xylocaine jelly. Then tube is inserted ventrally and medially, directing the nasal philtrum dorsally to facilitate passing the tube to the predetermined level. Humidified oxygen is to be administered at the flow rates of 50 to 100 ml/kg/min. In the absence of humidified oxygen, nonhumidified oxygen can be used in the emergency. But administration of nonhumidified oxygen causes dryness and irritation of the nasal mucosa leading to much discomfort to dogs and cats and intolerance to nasal cannula. Sometimes epistaxis may occur with the placement of nasopharyngeal or nasal cannula. This method can provide higher flow rate of supplemental oxygen.

- Nasal oxygen prongs- They are widely used in human patient. Their size varies from new born, pediatrics to adult. A prong is placed in each nares and aligned with the opening of the nostril. Prongs needs to be secured with sutures or surgical staples as it can slip out easily. In this method flow rate ranges from 3.0 L/min. to 6.0 L/min. (Mazzaferro, 2009).

- Nasotracheal oxygen administration- A red rubber catheter is passed through nostrils until it reaches proximal trachea. Catheter is then secured as in case of nasal cannula. This method provides 90% maximum fraction of inspired oxygen.
- Intubation- It is most suitable method of oxygen administration when 100% inspired oxygen administration is desired. This method requires anaesthesia, and intubation. This is not a common approach. It is generally reserved for cases with end stage laryngeal paralysis, laryngeal collapse or brachycephalic syndrome.

Oxygen Toxicity

- When excessive oxygen is administered over a prolong period, oxygen toxicity may occur.
- Excessive oxygen can cause alveolar damage and decrease pulmonary function.
- To avoid oxygen toxicity, small animals should receive a FiO_2 of < 60% for 1-2 days only.

Hyperbaric Oxygen Therapy (HBOT)

- In this therapy pure oxygen is used under pressure.
- In the 1940s, hyperbaric oxygen therapy became the standard therapy for the military divers in the United States of America.
- Its use in veterinary medicine is relatively new. Therapeutic use of HBOT in veterinary medicine dates back to 1998. Since then sporadic reports on the use of HBOT in animals have appeared in the literature. The HBOT is an emerging therapy in veterinary medicine.
- Hyperbaric chamber technology is used.
- HBOT is commonly used in humans to treat decompression sickness, or any sickness requiring high level of oxygen such as diving disorders, serious infections, and bubble of air in blood vessels, non-healing wound in diabetic patients or radiation injury. Other approved uses of hyperbaric therapy in USA are DCI, anemia due to severe blood loss, carbon monoxide poisoning, fire or heat burns, skin grafts, necrotizing soft tissue infections, osteomyelitis, acute traumatic ischemia, gas gangrene, and radiation injury.

- There are some unapproved use of hyperbaric oxygen therapy (HBOT) in humans in USA such as in asthma, Bell's palsy, depression, migraine, HIV, hepatitis, heart diseases, multiple sclerosis, stroke, Parkinson's disease, Alzheimer's disease. FDA (USA) has not approved the use of hyperbaric therapy in these conditions.
- In veterinary medicine, hyperbaric oxygen therapy is recommended in curing inflammation, poisonous bites from snakes, post-operative swellings, crushed injuries, burns, traumas (head, spinal cord, internal), pancreatitis, smoke inhalation, carbon monoxide poisoning, intervertebral disc disease, and nonhealing wound. It has also been recommended in the management of arthritis, infected wound in pets, aortic thromboembolism, gastric dilatation and volvulus, stroke and after cardiopulmonary cerebral resuscitation.
- The patient is placed safely and comfortably in a large chamber with 100% oxygen at 1.5 to 3 times that of normal atmospheric pressure. Treatment may last for 1-2 hours one to three times daily at four hours difference in two sessions. When lower pressure (1.5-2.0 ATA) is used treatment time lasts longer. At higher pressure (2.0-3.0 ATA) treatment time is shorter. Average treatment time at 2 ATA (atmospheres absolute) is 45 to 60 minutes (Clarke *et al.*, 2008).
- There are not much side effects. However, ear problem, blurry vision and barotraumas are seen in some cases.
- Primary mechanism of action of hyperbaric oxygen therapy (HBOT) are immediate hyper oxygenation of plasma and tissues, and decreased gas bubble size (air embolus). With hyperbaric oxygen therapy (HBOT), oxygen diffuses through tissues even when blood supply to the area is significantly compromised. Other beneficial effects of HBOT are an increase in leukocyte oxidative killing capacity (Babior, 1978; Allen *et al.,*1997), modulation of nitric oxide production (Tepper *et al.,* 2002; Thom, 2004; Fadini *et al.,* 2005), modification of growth factors and cytokine effect through regulation of their levels and/ or receptors (Bonomo *et al.*1998;Feng *et al.,*1998; Sheikh *et al.*, 2000), decrease in production of clostridial alpha toxin , synergism with fluroquinolones, amphotericin B, and aminoglycoside (Verklin *et al.,*1977;Adams and Mader,1987; Mader *et al.*, 1999; Hopf *et al.*, 2008), and decrease in vasogenic edema. High oxygen levels have bacteriostatic and bactericidal effects against gram positive, gram negative aerobic and anaerobic microorganisms (Gill and Bell, 2004).

- Delivery of 100% oxygen under pressure allows plasma to carry much more oxygen and reduces the need of hemoglobin based oxygen delivery (Boerema *et al.*, 1960).
- Hyperbaric oxygen therapy reduces inflammation by mediating cytokines, prostaglandins, and nitric oxide.
- Possible contraindications of HBOT described in humans are lung pathologies (pneumomediastinum, untreated pneumothorax), fever, animals having predisposition to seizures, animals with risk of aspiration (as in coma, unconsciousness), asthma and pregnancy.

Risk Involved in Hyperbaric Oxygen therapy

- Hyperbaric oxygen therapy is generally a safe procedure. Complications are rare. Nevertheless the therapy may carry some risk as described in human patients.
- Ear injuries due to change in air pressure.
- Temporary eye lens changes may cause temporary myopia.
- Barotrauma in lungs can cause lung collapse.
- Seizures may occur due to oxygen toxicity.
- Diabetic patients on insulin therapy may have risk of hypoglycemia.
- In certain circumstances there is a risk of fire due to oxygen rich environment of the treatment chamber.

References

Adams, K. and Mader, J. (1987). Aminoglycoside potentiation with adjunctive hyperbaric oxygen therapy in experimental *Pseudomonas aeruginosa* osteomyelitis [abstract 69]. Undersea and Hyperbaric Medical Society Annual Scientific Meeting; Baltimore, MD;1987.

Allen, D.B., Maguire, J.J., Mahdavian, M.,Wicke,C.,Marcocci,L., Scheuenstuhl, H., Chang, M., Le, A.X., Hopf, H.W. and Hunt, T.K. (1997). Wound hypoxia and acidosis limit neutrophil bacterial killing mechanisms. Arch. Surg. 132:991-996.

Babior, B.M. (1978). Oxygen-dependent microbial killing by phagocytes. N. Engl. J. Med. 198:659-668.

Boerema, I., Meijne, N.G., Brummelkamp, W.H., Bouma, S., Mensch, M.H., Kamermans, F., Hanf, M.S. and van Aalderen (1960). Life without blood. J .Cardiovasc Surg. 182:133-146.

Bonomo, S.R., Davidson, J.D., Yu, Y., Xia,Y.,Lin,X. and Mustoe, T.A., (1998). Hyperbaric oxygen as a signal transducer: upregulation of platelet derived growth factor-beta receptor in the presence of HBO2 and PDGH. Undersea Hyperb. Med. 25:211-216.

Boyle, J. (2012). Oxygen Therapy. In: Advanced Monitoring and Procedures for Small Animal Emergency and Critical Care. John Wiley and Sons Ltd., Chichester, U.K. doi: 10.1002/9781118997246. ch20.

Clarke, D. (2008). Effects of pressure. In: *Physiology and Medicine of Hyperbaric Oxygen Therapy.* Neuman,T.S. and Thom,S.R. (eds) Philadelphia, PA: Saunders; pp.513-526.

Fadini, G.P., Miorin, M., Facco, M., Bonamico, S., Barsso, I., grego, F., Menegolo, M., de Kreutzenberg, S.V., Tiengo, A., Agostini, C. and Avogaro, A.(2005). Circulating endothelial progenitor cells are reduced in peripheral vascular complications of type 2 diabetes mellitus. J.Am. Coll.Cardiol.45: 1449-1457.

Feng, J. J., Gibson, J.J., Constant, J.S., Hussain,M.Z.,and Hunt,T.K. (1998).Hyperoxia stimulates macrophage vascular endothelial growth factor (VEGF) production. Wound Repair Regen. 6:A252.

Ford,B.R. and Mazzaferro,M.E.(2006). Kirk and Bistner's Handbook of veterinary Procedures and Emergency Treatment. Saunders, Elsevier, Marrison, 8th edn. Gill, A.L. and Bell, C.N.A. (2004). Hyperbaric oxygen: its uses, mechanisms of actions and outcomes. Q. J .Med.; 97:385-395.

Hopf, H.W., Kelly, M. and Shapshak, D. (2008). Oxygen and the basic mechanisms of wound healing. In: *Physiology and Medicine of Hyperbaric Oxygen Therapy*. Neuman, T.S. and Thom, S.R. (eds.), Philadelphia, PA: Saunders; pp.203-228.

Loukopoulos,P. and Reynolds,W.(1996). Comparative evaluation of oxygen therapy techniques in anesthesized dogs: intranasal catheter and Elizabethan collar canopy. Austr.Vet. Pract. 26: 199-204.

Mader, J.T., Shirtliff, M.E., Bergquist, S.C. and Calhoun, J.(1999). Antimicrobial treatment of chronic osteomyelitis. Clin Orthop Relat Res 360:47-65.

Mazzaferro,E.M.(2015). Oxygen therapy. In:Small Animal Critical Care Medicine. Silverstein, D.C. and Hopper, K. (eds.), 2nd edn.,St.Louis,M.O.;Elsevier Saunders, pp 77-80.

Mazzaferro, E.M. (2009). Oxygen Therapy. In: Small Animal Critical Care Medicine. Silverstein, D.C. and Hopper, K. (eds.). W.B. Saunders, Philadelphia, pp.78-81.

Sheikh, A.Y., Gibson, J.J., Rollins, M.D., Hopf, H.W., Hussain, Z. and Hunt, T.K. (2000). Effect of hyperoxia on vascular endothelial growth factor levels in a wound model. Arch. Surg.135:1293-1297.

Tepper, O.M., Galiano, R.D., Capla, J.M., Kalka, C., Gagne, P.J., Jacobowitz, G.R., Levine, J.P. and Gurtner, G.C. (2002). Human endothelial progenitor cells from type II diabetics exhibit impaired proliferation, adhesion, and incorporation into vascular structures. *Circulation* 106:2781-2786.

Thom, S.R.(2004). Effects of hyperoxia on neutrophil adhesion. *Undersea* Hyperb Med .31 :123-131.

Verklin, R.M. Jr. and Mandell, G.L. (1977). Alteration of effectiveness of antibiotics by anaerobiosis. .J .Lab. Clin. Med. 89:65-71.

4

Anorexia in Heart Failure

Anorexia in dogs with heart failure is very important clinical sign of concern as continued anorexia leads to weight loss and cardiac cachexia. Cardiac cachexia in dogs is the main reason that contributes to an owners' decision to opt for euthanasia. It is therefore much more important to understand the various factors that lead to anorexia in dogs having heart failure. As per an estimate anorexia occurs in up to 84% dogs with congestive heart failure (Mallery *et al.*, 1999). Gastrointestinal (inappetance, anorexia, diarrhea) and respiratory (dyspnea, tachypnea, cough) embarrassment in heart failure cases are suggestive of cardiac decompensation.

Etiology

Various etiological factors (Fig. 26) have been described to contribute to the development of anorexia not only in dogs but also in humans with congestive heart failure.

- Many metabolic alterations take place in the body in dogs with heart failure. Recently, it has been observed that neurohormonal activation as a compensatory mechanism, increase in adverse free fatty acid metabolism, decrease in protective glucose metabolism and insulin resistance (in a few cases) in heart failure cases is responsible for detrimental changes in heart muscle leading to anorexia (Ashrafian *et al.*,2007). Suppression of myocardial energy metabolism (Lopaschuk, *et al.*, 2010), and myocardial tissue insulin resistance (Arkadievich, 2020) have also been observed in dogs with heart failure.
- Requirement of calories and nutrients in dogs with heart failure is high as compared to healthy ones due to decrease in blood flow and cardiac output. Failure to meet the increased nutritional demands may eventually lead to anorexia.
- Fatigue, nutritional deficiencies, alterations in ions, wasting and distress in cardiac compromised dogs complicates the condition adding to inappetance.

- Pulmonary edema and dyspnea owing to chronic venous congestion in heart failure cases causes gastrointestinal hypoxia leading to anorexia.
- Some cardiac medicines such as digoxin may have adverse central effect on appetite. Use of digoxin in some dogs may alter the taste leading to distraction from food.
- Administration of drugs with food is another important cause of resenting food in dogs owing to change in taste leading to anorexia.
- Dietary indiscretion and unpalatable diets are also important potential cause of anorexia.
- Release and increase in pro-inflammatory cytokines (IL-1, IL-6, TNF- alpha) in cases of heart failure contributes to increased energy requirement, anorexia, hypertrophy and fibrosis of cardiac muscles and negative inotropic effect.
- Gastrointestinal ulceration may develop as a result of azotemia and passive congestion. Ulcerations may lead to anorexia.
- Ascites in right heart failure may lead to passive gastrointestinal congestion adversely influencing the gut motility and absorption resulting into anorexia.
- Gut dysbiosis is being considered another important cause for anorexia in dogs with congestive heart failure. During last few years, compromised gut health and dysbiosis has attracted much attention in human medicine. Gut dysbiosis with an abundance of Proteobacteria (*Enteriobacteriaceae* and *Escherchia coli*) has been reported in dogs with heart failure (Seo et al., 2020). *Prevotella,* an unclassified species of *Catenibacterium, Clostridium cocleatum*, an unclassified species of *Erysepelotrichceae* and *Eubacterium biforme* are found in the gut of healthy dogs. So there a change in gut flora in the cases of heart failure in dogs. Some studies documented an increased concentration of trimethylamine N-oxide (TMAO) in dogs with congestive heart failure suggesting impaired gut health (Zeisel *et al.,* 2017, Janerio *et al*., 2018, Karlin *et al*., 2019). The increased TMAO concentrations in dogs with congestive heart failure has been ascribed to an increased abundance of *E. coli* (Karlin *et al*., 2019).
- Oxidative stress is also a potential cause for anorexia in dogs with compromised heart. Cell oxygenation is impaired in dogs with heart failure leading to dysfunction of cellular metabolism and production of large amount of free radicles. These free radicals cause further cellular

damages leading to anorexia. Supplementation of essential fatty acids in the food decreases the cellular sensitivity to free radicles and thus provides first protection against oxidative stress. Coenzyme Q10, vitamin E, and polyphenols are antioxidant molecules which can help cells to neutralize these free radicals.

- Deficiency of vitamins and selenium has attracted the attention of research workers in heart failure patients. Deficiency of thiamine, B12, vitamin E and selenium may cause or complicate heart disease. These deficiencies may not be the primary or sole cause of heart disease. But may be secondary to unbalanced diet. Some breeds may be predisposed to vitamin deficiency. It is a vicious cycle leading to anorexia. Thiamine deficiency can cause cardiomyopathy in humans. But role of vitamin B deficiency in the genesis of heart disease in dogs and cats remains to be investigated. Anorexia and urinary loss of water soluble vitamins can contribute to low concentration of vitamin B series in heart patients. Low thiamine level has been reported in humans with congestive heart failure (Seligmann *et al.,* 1991). Low level of vitamin B12 in cats with cardiomyopathy has been reported unrelated to the diet and the therapeutic use of furosemide (McMichael *et al.,*, 2000).
- Azotemia can be another factor in the genesis of anorexia in heart failure. Development of azotemia in heart failure cases is due to decreased cardiac output and secondary to renal involvement or progression of primary kidney disease. Reduced cardiac output is reflected by systemic hypotension, low venous oxygen tension, elevated blood lactate, increased BUN and creatinine and these parameters can be used as indicators of reduced cardiac output. Significant azotemia can lead to anorexia in dogs with congestive heart failure.
- Changes in liver enzymes have been observed in many dogs with heart failure. Use of antiarrhythmic drugs may also elevate liver enzymes. The elevation of liver enzymes may also lead to anorexia.
- Ionic changes have been recorded in dogs with heart failure. Due to diuretics there may be increased loss of potassium, chloride and sodium. Magnesium level may also vary depending on the disease severity and treatment .Hypomagnesaemia has been reported in dogs with congestive heart failure. So magnesium levels should also be monitored in dogs under the treatment of cardiac disease.
- Pain during eating of food due to associated severe periodontal disease in heart failure cases may resent dog to eat and thus contribute to anorexia..

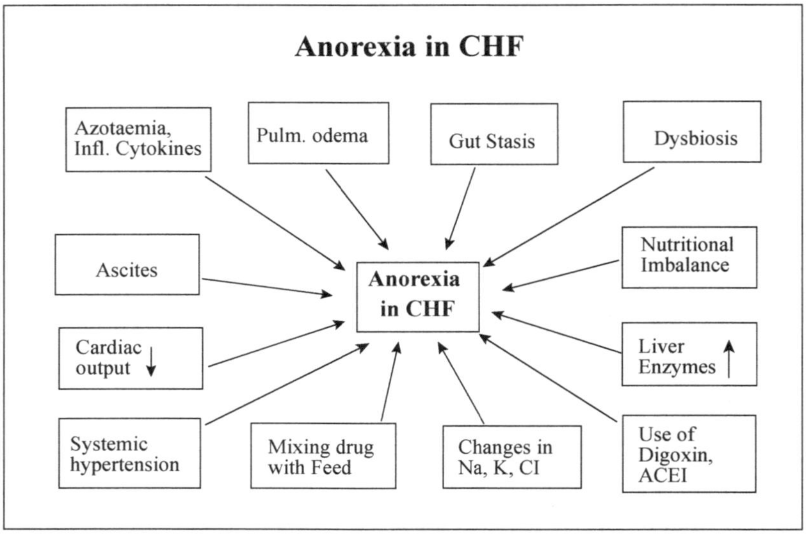

Fig. 26: Diagrammatic representation of various etiological factors responsible for anorexia in congestive heart failure in dogs.

Management of Anorexia

Management of anorexia in dogs with heart failure is not a simple issue. Resumption of food intake is very crucial for the wellbeing and longevity of the life of the patients. The following measures may be adopted to make the dog interested in food.

- Caloric Supplementation is very crucial for the survival of the patients.
- In early 60s protein restriction in dogs with congestive heart failure was adopted to decrease metabolic stress on liver and kidneys. But studies have not supported this thinking. On the other hand it is deleterious to restrict protein intake in dogs and cats with heart failure as it adds to cachexia. Unless there is severe renal failure, proteins in diet should not be restricted.
- Adding Carnitine, Taurine, Coenzyme Q10, Vit.E, Vit.B12 in the diet of dogs with heart failure is recommended. Most dogs with dilated cardiomyopathy do not have taurine deficiency. Dogs are able to synthesize adequate amount of taurine in the body. American cocker spaniels with dilated cardiomyopathy have shown lower level of taurine. Supplementation of taurine in the diet of dogs with dilated cardiomyopathy needs further research. However supplementation of taurine in cats with cardiomyopathy seems beneficial. Taurine may be

supplemented @ 500 mg PO BID/TID and Carnitine @ 1.0 g PO BID/TID.

- It is better not to mix drug with food to avoid resentment of the food by the dog owing to drug smell and /taste.
- Fish oil is a source of Omega-3 fatty acids (eicosapentaenic acid-EPA and docosahexaenoic acid-DHA) and has shown benefits in decreasing cytokine production and improve endothelial function in chronic inflammation associated with congestive heart failure. Therefore addition of EPA and DHA @ 40 mg/kg and 25 mg/kg respectively can be adopted to improve appetite.
- There is a controversy on the use of taurine supplementation in dogs as dogs are able to synthesize taurine.
- L-carnitine deficiency has been associated with DCM and its supplementation has shown beneficial effect. But in most cases, the effect of supplementation is minimal.
- Coenzyme Q10 is an anti-oxidant and it is believed to be cardio protective. Supporting data in this regard is not enough.
- Plain rice, honey, cooked eggs, fresh veg., yoghurt, watermelon, musk melon, orange, banana may be tried as an alternate food . Homemade diets are palatable and owners can have control over the ingredients.
- Sodium level can be easily restricted in these homemade diets. Carnitine, taurine, amino acids, vitamin B12, magnesium can be easily added as per need. Potassium level may be kept as per requirement of the individual case.
- Fresh and hot food in small volume given frequently may encourage the dog to eat.
- Garlic powder may be sprinkled over the feed to increase palatability.
- A small quantity of very palatable human food (meat or gravy without salt or soup with potassium chloride) may be mixed in dog food to improve taste.
- Anabolic steroids, glucocorticoids, benzodiazepines, cyproheptadine, mirtazapine, megesterol acetate and vitamin B complex are used in general as orexogenic drug. However, specific recommendation regarding their use as orexogenic drug in dogs with congestive heart failure is lacking.

- Anabolic steroids antagonize the catabolic effect of illness, trauma and aging. They may stimulate hematopoiesis, appetite, and weight gain. These drugs may cause hepatotoxicity, masculinization, and early closure of bony epiphyses in young animals . Anabolic steroids should not be used in dogs and cats with congestive heart failure because of sodium and water retention.
- Benzodiazepines (BZD) have a good reputation of appetite stimulation effects independent of their anxiolytic activity. Their binding to GABA- A receptors produces a strong dose-dependent increase in food consumption without affecting hunger level and degree of satiety. These drugs act specifically to enhance taste and other sensory characteristics of food. Intravenous administration of diazepam is an appetite stimulant to cats.. Oxazepam, a metabolite of diazepam, can be given orally to cats but diazepam is the more effective appetite stimulant with a sedative effect.
- Cyproheptadine is an antihistamine with serotonin-antagonist action being prescribed as an appetite stimulant in cats . It is a 5-HT2 receptor antagonist. Endogenous opiates, which stimulate eating, are normally excreted by the lateral hypothalamus. Their release is inhibited by serotonin and cholecystokinin release, thus causing inappetence. Cats are very sensitive to changes in serotonin concentrations, so serotonin antagonists are very potent in cats. CNS excitement and aggressive behavior may occur in some cats on cyproheptadine therapy.
- Mirtazapine is an antidepressant being used in humans to treat moderate to severe depression. Mirtazapine is not a serotonin or norepinephrine reuptake inhibitor .It is an antagonist of presynaptic α2-adrenergic autoreceptors and heteroreceptors on both norepinephrine and serotonin (5-HT) presynaptic axons, plus is a potent antagonist of postsynaptic 5-HT2 and 5-HT3 receptors. Antihistaminic drowsiness is a common effect. It is used in the treatment of GI disease or liver or kidney disease. Mirtazapine can also be used to alleviate the nausea and appetite loss that accompanies chemotherapy. It is an effective appetite stimulant and antiemetic for cats with chronic kidney disease. It is given once a day to dogs and twice a week in cats with severe liver or kidney disease. Since liver and kidneys are compromised in heart failure, mirtazapine may be tried as appetite stimulant. Information about the use of mirtazapine as an appetite stimulant in dogs with heart failure is not sufficient.
- Megestrol acetate is a synthetic progestin. It has significant antiestrogen and glucocorticoid activity leading to adrenal suppression. It has been

used as appetite stimulant in people with cancer to deal cachexia. It is expected to have a similar effect in anorectic cats and dogs. Its use in animals with pregnancy, uterine disease, diabetes mellitus, or mammary neoplasia is not recommended. A profound adrenocortical suppression, adrenal atrophy, and diabetes mellitus, may occur in cats with the use of megestrol acetate.

- Vitamin B complex is another commonly used orexogenic drug. Its preparations can be used oral or parenteral
- Out of sympathy owners offer foods such as cream, ice cream, pizza, canned and fried food to their heart compromised dogs. Actually these foods are detrimental to cardiac compromised dogs and deteriorate the condition further. Therefore these food items should not be fed to dogs with heart failure.

References

Arkadievich,O.D.(2020).Metabolic markers of myocardial insulin resistance in dogs with heart failure. Open Vet.J. 10:363-370.

Ashrafian,H.,Frenneaux,M.P. and Opie,L.H. (2007). Metabolic mechanisms in heart failure. Circulation116:434-448. https://doi.org/10.1161/Circulation AHA.107.702795.

Janeiro, M. H., Ramirez, M. J., Milagro, F. I., Martinez, J. A. and Solas, M.(2018). Implication of trimethylamine N-oxide (TMAO) in disease: potential biomarker or new therapeutic target. Nutrients 10: 1398

Karlin, E. T., Rush, J. E. and Freeman, L. M. A.(2019).Pilot study investigating circulating trimethylamine N-oxide and its precursors in dogs with degenerative mitral valve disease with or without congestive heart failure. J. Vet. Intern. Med .33: 46–53.

Lopaschuk, G.D., Ussher, J.R., Folmes, C.D.S., Jaswal, J.S. and Stanley, W.C. (2010). Myocardial fatty acid metabolism in health and diseases. Physiological Reviews 90. https://doi.org./10.1152/physrev.00015.2009.

Mallery, K.F., Freeman, L.k., Harpster, N.K. and Rush, J.E. (1999). Factors contributing to the euthanasia decision in dogs with congestive heart failure. Am.J. Vet.med. Assoc. 214:1201-1204.

McMichael, M.A., Freeman, L.M., Selhub, J., Rozansk, E.A., Brown, D.J., Nadeau, M.R. and Rush, J.E. (2000). Plasma homocysteine, B vitamins and amino acid concentrations in cats with cardiomyopathy and arterial thromboembolism. J. Vet.Intern. Med. 14:507-512.

Seligmann, H., Halkin, H., Rauchfleisch, S., Kaufmann, N., Tal, R., Motro, M., Vered, Z.V.I. and Ezra, D. (1991). Thiamine deficiency in patients with congestive heart failure receiving long term furosemide therapy: a pilot study. Am. J. Med. 91:151-155.

Seo, J., Matthewman, L., Xia, D., Wilshaw, J., Yu-Mei Chang and Connolly, D.J. (2020). The gut microbiome in dogs with congestive heart failure: a pilot study. Sci.rep. 10:13777. https://doi.org/10.1038/s41598-020-70826-0.

Zeisel, S.H. and Warrier, M. (2017). Trimethylamine N-oxide, the microbiome, and heart and kidney disease. Annu.Rev.Nutr.37:157-181.

5

Dietary Considerations in Heart Failure

Diet plays an important role in the management of diseases. It is very much true in the management of dogs with heart failure also. Managing the diet of a dog with heart failure is depending on the severity and stage of heart failure and presence / absence of concurrent diseases as well deficiencies. Till now emphasis has been given on the severe restriction of salt (sodium) in the diet of the dogs with heart failure because body is unable to excrete sodium and water. But current concepts are emphasizing on the role of other nutrients as well and keeping salt level normal in the diet as severe restriction of sodium chloride in the diet also stimulate RAAS activation. Salt restricted diets are poorly palatable and may result in inadequate caloric intake. Heart failure itself is a calorie reducing state and poorly palatable diet (salt restricted diet) further aggravates the caloric deficiency leading towards cachexia.

Important Considerations

- Optimum weight should be maintained.
- Cachexia should be dealt with. Anorexia should be managed properly so that it does not lead to cachexia.
- Appetite may be stimulated with supplementation of omega -3 polyunsaturated fatty acids. Fish oil can be used as a source of omega -3 as it reduces inflammatory cytokines in dogs suffering from heart failure.
- Association of obesity with heart diseases in dogs is not well studied. Nevertheless, obesity may cause decrease in cardiac output, change in blood pressure, change in heart rate, and neurohumoral activation. Exercise is restricted in dogs with heart failure, therefore obesity reduction in obese dogs having heart failure is attained through caloric restriction.
- Severe dietary restriction of sodium can also activate RAAS. It is doubtful whether salt restriction during an early stage of heart failure has any advantage. It has been reported that starting a low sodium diet

before the onset of heart failure may only serve to activate RAAS system and may speed the onset of overt heart failure. Hyperkalaemia has been observed in dogs affected with heart failure maintained on furosemide along with salt restriction (Roudebush *et al.*,1994).Dogs with an early heart failure should be given sodium < 80 mg/100 kcal (0.1-0.35%DM basis). Sodium content of 50 to 80 mg/100 kcal can be considered for dogs with grade B2 and C heart failure. Dog with grade D heart failure are to be hospitalized and maintained on parenteral alimentation. Diet should have adequate calories and protein. The situation becomes complicated because of in appetence in cases of heart failure. Some efforts can be made to enhance the flavor of the food by warming , by adding very small amount of very palatable human foods (meat or gravy without salt or soup with potassium chloride rather than common salt), or adding garlic powder. Hand feeding may be tried. Ailing dogs /cats may be provided with small quantity of food several times a day.

- Treats and human food having high contents of salt should be avoided.
- Dogs on Angiotensin converting enzyme inhibitors (ACE-I) and spironolactone (aldosterone inhibitor and potassium sparing diuretic) therapy may develop hyperkalaemia. Even some commercial food having higher contents of potassium may lead to hyperkalaemia.
- Deficiency of vitamin B1 (thiamine), vitamin E, selenium, and taurine may secondarily complicate heart disease.
- Once restriction of protein in the diet of dogs with heart diseases was considered to reduce metabolic stress on liver and kidneys. But there is no scientific evidence to support this thinking. On the contrary protein restriction may be deleterious as such dogs may be predisposed to loss of lean body mass and may add to cachexia. Feeding of high quality protein to dogs with heart failure may be considered unless there is severe renal compromise.
- Dogs are able to synthesize sufficient amount of taurine. In most of the dogs with DCM taurine level is not deficient. Nevertheless, some breeds of dogs with DCM (American Cocker Spaniel, Golden retriever, Labrador retriever, Newfoundland, Dalmatian, Portuguese water dogs, Bull dogs) may show insufficient level of taurine possibly due to some dietary factors (high fiber diets; boutique pet food containing peas, lentil, legume seeds or potatoes; lamb and rice based diets). Occasionally dogs maintained on vegetarian or vegan diet may have taurine deficiency and can make them susceptible to heart diseases. Some effect of taurine

supplementation in dogs with heart disease has been observed but it is not very conspicuous as seen in cats. Beneficial effect of supplementation of taurine in dogs with heart disease, whatsoever, may possibly be due to its positive inotropic effect. Taurine is an essential nutrient for cats. Its prolonged deficiency in cats may lead to myocardial failure. Promising results have been observed when deficient cats are supplemented with 250 to 500 mg taurine twice daily in their food. Deficient dogs may be supplemented with 500 mg taurine twice daily.

- Supplementation of arginine in the diet of humans with congestive heart failure has been shown to improve endothelial dysfunction without having adverse effect on myocardial contractility.
- Studies have shown decreased levels of n-3 polyunsaturated fatty acids, eicosapentaenoic acid (EPA) and docosahexaenoic acid (DHA) in dogs with congestive heart failure. Therefore supplementation of fish oil in the diet of dogs with heart failure may be beneficial (Freeman *et al.*, 1998).
- It has been shown that thiamine deficiency can cause cardiomyopathy in humans (Seligmann *et al.*,1991). But such relationship is still to be proved in dogs and cats. Lower levels of water soluble vitamins could be ascribed to anorexia and urinary loss.
- Cats with cardiomyopathy have shown significantly lower level of vitamin B12 (McMichael *et al.*, 2000). Supplementation of vitamin B12 in cats with cardiomyopathy may be beneficial.
- L-carnitine is reported to be associated with primary cardiomyopathy in animals . A good beneficial response of carnitine supplementation has been reported in some boxers and American cocker spaniels affected with dilated cardiomyopathy ((Keene *et al*, 1991; Kittleson *et al.*, 1997). Supplementation of L-carnitine may be given to dogs @ 50-100 .0 mg/ kg b.wt. thrice daily or 1.0-2.0 g twice or thrice daily. It is reported that the beneficial effect of L-carnitine may be due to improving myocardial energy production.
- Some cardiac drugs may cause depletion of magnesium. Therefore monitoring of blood magnesium level may be necessary in dogs receiving cardiac drugs. Low magneseium level increases the risk of arrhythmias. Dogs on cardiac drugs showing low level of magnesium need magnesium supplementation

- Supplementation of Coenzyme Q 10 in the diet of dogs with heart failure, a cofactor and anti-oxidant, may have some beneficial effect for its anti-oxidant property and improving myocardial metabolic efficiency. Well controlled studies are still lacking.
- Increased level of biomarkers of oxidative stress and a decrease in vitamin E (an anti-oxidant) have been observed in dogs with heart failure . Hence addition of antioxidants in the diet can be beneficial.

Recommended Home Made Food

- Boiled rice, honey, boiled eggs, lean home cooked meats or cereals (all low salt) and fresh vegetables and fruits (carrots, green beans, apple, orange, banana) are recommended.

Prohibited Foods

- Milk, milk products, ice cream, cream, butter, cheese, margarine.
- Chocolates
- Bread, pizza, pretzels and other munchies.
- Sauce, ketchup
- Potato chips, popcorn, crackers and other snacks used by humans
- Commercial available dog treats
- Processed food such as processed meats, liver, kidneys, canned fish
- Canned vegetables

Prescription Diets

- Some prescription diets are available for dogs with congestive heart failure. These diets are having controlled levels of sodium and chloride.

References

Freeman, L.M., Rush, J.E., Kehayias, J.J., Ross, J.N., Meydani, S.N., Brown, D.J., Dolnikowski, G.G., Marmor, B.N., White, M.E., Dinarello, C.A. and Roubenoff, R. (1998). Nutritional alterations and the effect of fish oil supplementation in dogs with heart failure, J. Vet. Intern. Med.12: 440-448

Keene, B.W., Panciera, D.P., Atkins, C.E., Regitz, V., Schmidt, M.J. and Shug, A.L. (1991). Myocardial L-carnitine deficiency in a family of dogs with dilated Cardiomyopathy. J. Am. Vet. Med. Assoc.198: 647-650.

Kittleson, M.D., Keene, B., Pion, P.D. and Loyer, C.G. (1997). Results of the multicenter spaniel trial (MUST): taurine and carnitine-responsive dilated cardiomyopathy in American cocker spaniels with decreased plasma taurine Concentration. J.Vet.Intern. Med.11: 204-211.

McMichael, M.A., Freeman, L.M., Selhub, J., Rozanski, E.A., Brown, D.J., Nadeau, M.R. and Rush, J.E. (2000). Plasma homocysteine, B vitamins, and amino acid concentrations in cats with cardiomyopathy and arterial thromboembolism. J. Vet. Intern. Med. 14: 507-512.

Roudebush, P., Allen, T.A., Kuehn, N.F., Magerkurth, J.H. and Bowers, T.L. (1994). The effect of combined therapy with captopril, furosemide, and a sodium-restricted diet on serum electrolyte concentrations and renal function in normal dogs and dogs with congestive heart failure. J.Vet.Intern.Med. 8 : 337-342.

Seligmann, H., Halkin, H., Rauchfleisch, S., Kaufmann ,N., Tal, R., Motro, M., Vered, Z.V.I. and Ezra, D. (1991). Thiamine deficiency in patients with congestive heart failure receiving long-term furosemide therapy: a pilot study . Am.J.Med. 91: 151-155.

6

Cardiac Arrest

The word cardiac arrest (CA) or heart attack is frightening. In animals heart attack does not occur in the manner as it occurs in humans. In humans the main cause of cardiac arrest or heart attack is coronary infarct. The prevalence of cardiac arrest or cardiopulmonary arrest (CPA) or heart attack in animals is relatively very low as coronary infarct is not a common disease in animals. In animals cardiac arrest or cardiopulmonary arrest or circulatory arrest is seen when cardio-pulmonary systems cease to function. If a dog fails to breathe for few minutes (may be 5-6 minutes), it may lead to cardiac arrest. The process of cardiac arrest is rapid leading to collapse, unconsciousness, and ceasing of respiration.

Etiology

- Trauma leading to severe blood loss.
- Pneumothorax
- Brain damage due to head trauma.
- Cardiac diseases with heart failure.
- Pericardial tamponade
- Tumors on or around heart.
- Severe unstable cardiac arrhythmias such as ventricular tachycardia, 3rd degree heart blocks.
- Exposure to toxin (plant , food , chemicals or some drugs)
- Snake bites.
- Untreated cases of heartworm.
- Anesthetic complications.
- Electrocution or chewing electric cords.
- Severe hypoxia
- Severe anemia

- Severe electrolyte imbalance (hyperkalemia, hypocalcaemia, hypomagnesaemia)
- Severe dehydration
- Shock
- Severe toxemia
- Vagal stimulation

Predisposition

- There is no breed, age or sex predisposition to cardiac arrest.
- It can occur in any breed, sex or age group.

Clinical manifestations

If breathing ceases for more than 5-6 minutes, cardiac arrest may take place. Its predominant signs are as follow.

- Fixed and dilated pupil
- Change in color of mucus membrane /Cyanosis (due to diminished oxygen in the blood)
- Gasping/dyspnea/ open mouth breathing
- Low body temperature (hypothermia)
- Poor or no response to stimulation
- Marked hypotension
- Weak Pulse and inaudible heart sound
- Syncope (spontaneous unconsciousness)/collapse

Diagnosis

- Assessment of the dogs with respect to responsiveness, breathing pattern and pulse is very important with suspected cardiopulmonary arrest to undertake cardiopulmonary cerebral resuscitation (CPCR) measures.
- Thorough clinical examination
- Blood pressure and pulse monitoring
- Oxymetry
- Chest X-ray

- CBC
- Biochemistry profile
- Urine analysis
- Blood gas analysis
- Echo and Electrocardiographic examinations.

Absence of palpable pulse does not always suggest CPA. In humans the diagnosis of CPA by pulse palpation is not always reliable. Electrocardiography or Doppler blood pressure monitoring in unresponsive apneic cases has also not been recommended to diagnose CPA. However, it can be of advantage to diagnose impending CPA cases. End tidal carbon dioxide (EtCO2) is recommended for intubated cases at risk of CPA.

Treatment

- Rapid assessment of the case and immediate cardiopulmonary resuscitation (CPR) is the key of managing cardiopulmonary arrest (CPA).Mortality rate in cases of cardiopulmonary arrest is very high despite treatment. If a dog or cat is apneic or unresponsive, immediate CPR should be initiated.
- Cardiopulmonary cerebral resuscitation (CPCR) measures are initiated in three stages viz.basic life support, advanced life support and post resuscitation care (Pablo, 2003).

(a) Basic Life Support (BLS)- Basic life support measures consist of performing chest compression, assisting ventilatory breath and defibrillation for ventricular fibrillations. These measures are popularly called ABCs. A stands for airways, B stands for breathing and C stands for Circulation.

Extending the neck straight and pulling the tongue forward can assist in cleaning the upper airway. Suction may be applied if deemed necessary. Foreign material or vomit in the mouth should be cleaned. If airways obstruction is not cleared, emergency tracheotomy should be done.

Dogs and cats are given 2 breaths 1-2 second in duration with positive pressure ventilation using 100% inspired oxygen. After clearing airways endotracheal tube should be placed and positive pressure ventilation with 100% oxygen should be started. Heart and femoral pulse should be monitored. Use of Doxapram, a respiratory stimulant, to improve breathing in cases of CPA may be deleterious as it decreases cerebral blood flow, and increases cerebral oxygen consumption and requirement (Roll and Horsch, 2004).

If heat beat is not audible, chest compression over the heart should be initiated at the rate of 80-120 compressions/minute (Plunkett and McMichael., 2008) for dogs and cats in lateral recumbancy. Depth of chest compression should be around half the width of chest. Leaning on the patient should be avoided. CPR should be performed in 2 minute cycle without interruption. Open chest cardiac massage has been recommended in all cases of chest trauma, and/or hemorrhage (haemothorax, pneumothorax, cardiac tamponade).

(b) Advance Life Support (ALS) - It includes venous access, electrocardiographic assessment of heart, drug administration and defibrillation as case may be. After accessing vein aggressive fluid therapy should be given to deal hypovolemia. Care should be exercised to avoid overzealous fluid therapy if there is normal fluid volume. Hypertonic saline, crystalloids, colloids, blood products, and hemoglobin based oxygen carrying solutions are used for fluid resuscitation. In initial stage if placement of intravenous catheter fails, emergency drugs (lidocaine, epinephrine, atropine, naloxone etc.) can be administered through endotracheal tube (Emmerman *et al.*, 1988). Naloxone (0.02-0.04 mg/kg IV; 0.04-0.1 mg/kg IT) is recommended in case of opioid toxicity. Lidocaine @2.0-4.0 mg/kg IV/IO or 4.0-10.mg/kg IT is used in dogs with ventricular arrhythmias in CPR. In cats lidocaine is recommended @ 0.2 mg/kg IV/IO/IT but should be used very cautiously. Their doses can be increased 2to 3 fold. If closed circuit CPCR is being followed, intracardiac injections should be avoided. After establishing the type of arrhythmia, drug administration and defibrillation is to be undertaken. Vasopressor therapy, vagolytic therapy, electrical cardio version and correction or reversion of conditions led to CPA are to be undertaken. Aim of vasopressor therapy is to improve coronary and cerebral blood flow. Epinephrine (0.01 mg/kg every 3-5 min IV/IO; 0.03-0.1 mg/kg IT) is the commonly used drug for this purpose. The drawback is that the epinephrine increases myocardial oxygen demand. Vasopressin does not increase myocardial oxygen demand and can be used as a substitute @ 0.2- 0.8U/kg IV/IO or 0.4- 1.2 U/kg IT

Vagolytic therapy is used in cases of CPA related to increased vagal tone or asystole.

Atropine is generally used vagolytic drug @ <0.04 mg/kg IV/IO over 10 minutes or 0.08 mg/kg Intra tracheal (IT). Electrical cardioversion (defibrillation) is needed in case of ventricular fibrillation, and pulse less ventricular tachycardia. Biphasic defibrillator seems better.

(c) Post resuscitation care (PRC) - Some cases restored to normal cardiac rhythm with advanced life support may not recover if original cause of CPA has not been addressed.

Therefore resuscitated cases need intensive monitoring and substantial cardiovascular and ventilatory supports following CPA. Mild hypothermia after resuscitation from CPA seems beneficial. Hyperglycemia in post resuscitated humans is associated with worse neurological outcome (Steingrub and Mundt, 1996).Therefore glucose administration during CPR is not advisable until unless there if hypoglycemia.

Brain injury, disseminated intravascular coagulopathy, gut reperfusion syndrome and renal failure may be precipitated in poorly perfused animals during CPA. Therefore adequate aggressive supportive therapy and intensive monitoring is needed to optimize blood pressure, cardiac output, oxygenation, ventilation, and perfusion of vital organs (Steingrub and Mundt, 1996).

Monitoring During CPR

- Electrocardiogram
- End-tidal carbon dioxide
- Blood gas and electrolyte analysis

Post Cardiopulmonary Arrest Syndrome

- It is a complex pathological response observed as a sequel to global body ischemia following reperfusion. It is commonly seen in humans.
- Its major components are brain injury, myocardial dysfunction, systemic ischemia/ reperfusion injury and pre-existing comorbidities.
- Ischemia causes initial tissue insult. Return of perfusion adds to further damage to the organ and to distant organs also. There is cellular swelling due to influx of intracellular water and liberation of highly reactive superoxide anion (O2-) and uric acid as byproduct of reaction in the cells leading to production and systemic release of proinflammatory eicosanoids, disruption of cell membrane permeability and eventually cell death.
- Systemic ischemic reperfusion injury is a sepsis like syndrome because of common characteristics in relation to coagulation and inflammation.
- The goal of therapy is to sustain spontaneous circulation and perfusion of most vital organs (brain and heart). There is dearth of studies in dogs or cats to recommend specific treatment protocol for post cardiopulmonary arrest syndrome. Based on limited studies and human studies hemodynamic optimization protocol may be followed.

- Vasopressors (norepinephrine and dobutamine) are the first-line treatment for shock.
- Oxygen saturation between 94 % and 98 % is to be maintained to avoid fatality.
- Therapeutic mild hypothermia helps to resuscitated patients. Strict temperature control is required.
- Electrocardiogram, oxymetry, capnography, and electroencephalography (EEG) along with blood pressure, temperature, and vital signs are to be monitored frequently.
- Evidence is lacking to support the administration of corticosteroids, anti-seizure prophylaxis, mannitol or metabolic protectants after cardiac arrest. Patients showing seizure should be appropriately treated. Glycemic control and metabolic management yields good neurological outcome. Hypertonic saline may be administered in animals suspected of post cardiopulmonary arrest syndrome
- Recovery from acute kidney injury is essential for survival and a good neurological outcome.

References

Emmerman, C.L., Pinchak, A.C.,Hancock, D. and Hagon,J.F. (1988). Effect of injection site on circulation time during cardiac arrest. Crt.Care Med.16:1138-1141.

Pablo,L.S.(2003). Current concepts in cardiopulmonary resuscitation. World Small Anim.Vet. Assoc. World Congr. Proc. 2003.

Plunkett,S.J. and McMichael,M. (2008). Cardiopulmonary resuscitation in small animal medicine: an update. J.Vet.Intern.Med. 22:9-25.

Roll,C.,and Horsch,S. (2004). Effect of doxapram on cerebral blood flow velocity in pre term infants. Neuripediatrics 35:126-129.

Steingrub,J.S. and Mundt,D.J.(1996). Blood glucose and neurological outcome with global brain ischaemia. Crit.Care Med.24:802-806.

7

Shock

Shock is a sequel to trauma or severe diseases such as heart failure, sepsis or acute inflammatory conditions like pancreatitis. Shock is generally associated with decreased supply and utilization of oxygen. The term shock *is* a chain of physiological events that can result in death if immediate rational treatment is not instituted.

The shock can be appropriately defined as decreased cellular metabolism (inadequate energy production and /or decreased cellular oxygenation) owing to decreased blood flow that leads to cell death and organ failure (Hopper *et al.*,2012).

Etiology

- The main cause of the shock is acute peripheral circulatory failure leading to tissue hypoperfusion and decreased oxygenation.
- In untreated cases, hypoxia and anaerobic cellular metabolism may eventually cause cell death.
- In animals, hemorrhages, heart disease, sepsis, trauma that adversely affects circulation and cellular oxygenation may cause shock.

Pathogenesis

- Decreased circulatory volume causes release of epinephrine from adrenal medulla resulting in an increase in heart rate, increase myocardial contractility and peripheral vasoconstriction.
- Because of vasoconstriction, systemic vascular resistance is increased but perfusion to skin, muscles and gastrointestinal tract is decreased as blood is diverted to vital organs (brain and heart).
- Baroreceptors stimulate neural activity increasing the level of catecholamine. The increased level of catecholamine triggers the release of rennin from the kidneys initiating rennin-angiotensin aldosterone system (RAAS) causing increased resorption of water and sodium.

- With the help of angiotensin converting enzyme (ACE), angiotensin I (AT-I) is converted in to angiotensin II(AT-II) causing further vasoconstriction and release of norepinephrine leading to further increase in heart rate and myocardial contractility.
- Redistribution of blood flow as per priority of the body causes decrease in tissue oxygenation in many organs resulting in anaerobic metabolism leading to dilation of precapillary vessels , blood pooling within venules, lactic acidosis, increased intracellular concentration of sodium and water due to decrease in ATP and eventually causing the loss of cellular integrity.
- Decreased blood flow to gastrointestinal tract causes gut hypoxia damaging intestinal endothelium and disrupting the integrity of gut mucosal barrier. As a result of bacterial or endotoxin translocation, cytokines are released that in turn causes multiorgan dysfunction or failure.

Stages of Shock

The shock is a progressive condition in untreated cases. Broadly it can be categorized in two stages as compensatory shock and decompensatory shock. The later one can be further categorized as an early and late decompensatory (terminal) stage.

Compensatory Stage of Shock

- In this stage epinephrine release is the predominant event. Increase in heart rate and myocardial contractility is the effort of the body to increase tissue perfusion and oxygenation.
- Retention of water and sodium by the kidneys under the influence of aldosterone; and movement of fluid from the interstitial space to intravascular space under the influence of other neural changes are other important efforts of the body to increase circulatory volume leading to hypermetabolic state at cellular level (increasing energy and oxygen requirements).
- To meet the increasing demand, adrenocorticotropic hormone (ACTH), cortisol and glucagon are released to metabolize energy. Clinical signs observed during compensatory stage of shock are mild tachycardia, injected mucus membranes, rapid capillary refill time, and normal to slightly elevated blood pressure.

- The aim of treatment in this stage is to re-establish effective circulatory volume and to arrest the progression of shock to decompensatory stage.

Decompensatory Stage of Shock

- If the underlying cause of shock is not resolved, body's efforts to cope up decreased circulatory volume and hypoperfusion of tissues during compensatory stage remain futile to check the progression of shock and compensatory stage turns into decompensatory stage
- Decompensatory stage has two sub stages as an early and late stage.

Early Decompensatory Stage

- During early decompensatory stage blood flow is redistributed as peripheral vasoconstriction reduces blood supply to skin, muscles, liver, kidneys and gastrointestinal tract so that an adequate perfusion to the brain and heart can be maintained.
- Hypoxia in the areas of reduced blood supply owing to vasoconstriction initiates anaerobic metabolism with a concurrent increase in lactic acidosis resulting into loss of cellular integrity and in turn increases the production and release of prostaglandins and other chemical mediators of inflammation.
- Gut hypo perfusion damages the gut mucosal barrier allowing bacterial translocation into the systemic circulation.
- Renal hypo perfusion results in oliguria and tubular necrosis.
- Myocardial contractility is reduced and arrhythmias may develop.
- Events during the early decompensatory stage of shock are clinically manifested as pale mucus membrane, tachycardia,, prolonged refill time, hypotension, weak thread pulse, low body temperature, depression, and dullness.
- With the persisting hypo perfusion and tissue hypoxia, the body's regulatory and compensatory mechanisms starts failing and early stage of decompensatory shock progresses in to late decompensatory shock.
- Autorgulatory escape causes bradycardia and massive vasodilation within organs leading to pooling of blood and further decrease in circulatory blood volume and tissue oxygenation. At this stage blood circulation to heart and brain is also compromised resulting into collapse.

Late Decompensatory or Terminal Stage of Shock

- The terminal stage of shock or late decompensatory shock is clinically characterized by pale or cyanotic mucus membranes with no capillary refill, reduced heart rate, hypotension, weak/absent pulse, hypothermia, oliguria/anuria, mental depression, stupor or coma, lung edema, and cardiac arrest.
- Response to aggressive fluid therapy and other supportive therapeutic interventions is generally remain unrewarding in the late stage of decompensatory shock.

Abstract Clinical Findings in Different stages of shock in Dogs

Compensatory Shock

- Rectal temperature normal to slightly low.
- Tachycardia
- Pale mucus membrane.
- Hyperemic mucus membranes in distributive shock.
- Capillary refill time normal to slightly prolonged. It is rapid in cases of distributive shock.
- Tachypnea.
- Normal to slightly low blood pressure.
- Mental alertness

Early Decompensatory Shock

- Rectal temperature slight to moderately low.
- Moderate tachycardia
- Pale or white mucus membranes.
- Prolonged capillary refill time.
- Severe tachypnea.
- Mild to moderately low blood pressure.
- Mental alertness reduced

Late Decompensatory Shock

- Rectal temperature moderate to markedly low
- Mild tachycardia to bradycardia
- Muddy mucus membranes
- Prolonged capillary refill time
    - Reduced /slow respiration rate
- Marked hypotension unresponsive to fluid therapy
- Mental alertness markedly reduced.

Classification of Shock

Shock is classified into different categories based upon initiating event (Fig. 27). There are three main categories of shock as circulatory, metabolic and hypoxic shock (Hopper *et al.,* 2012). Circulatory shock is further divided into three subcategories : hypovolemic, cardiogenic and distributive (Boulpaep, 2009).

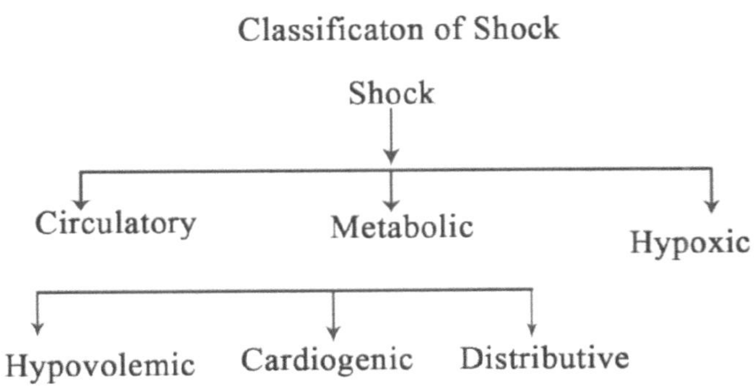

Fig. 27: Classification of shock

Hypovolemic shock

- This type of shock is due to a significant decrease in effective circulatory volume as seen in cases of severe dehydration (Fig. 28) either due to excessive urination, vomiting, diarrhea, and/or acute hemorrhages.
- Clinically such cases progress through all stages of shock. Pale mucus membranes, prolonged capillary refill time, tachycardia and weak

pulse are classical signs of hypovolemic shock that actually represent decompensatory stage.

- Aggressive fluid replacement therapy is the key to the treatment of hypovolemic shock. Crystalloids, artificial colloids, blood and it products, hypertonic saline or their combinations are recommended.
- Crystalloids (80-90 ml/kg in dogs, 40-60 ml/kg in cats divided in four doses giving each over 15 minutes with reassessing the dog/cat) are more effective in the treatment of shock associated with dehydration.
- Artificial colloids (10.0-20.0 ml/kg/day) are better choice in cases of third spacing resulting from hypoproteinemia.
- Hypovolemic shock arising from severe hemorrhages can be better treated with whole blood, plasma or packed red cells.
- Treatment of hemorrhagic shock due to rodenticides requires fresh / frozen plasma or fresh whole blood.
- Bleeding needs to be controlled by the use of pressure bandage or as the case may be.
- Rupture of spleen or liver needs immediate surgical intervention. Oxygen therapy may also be instituted.

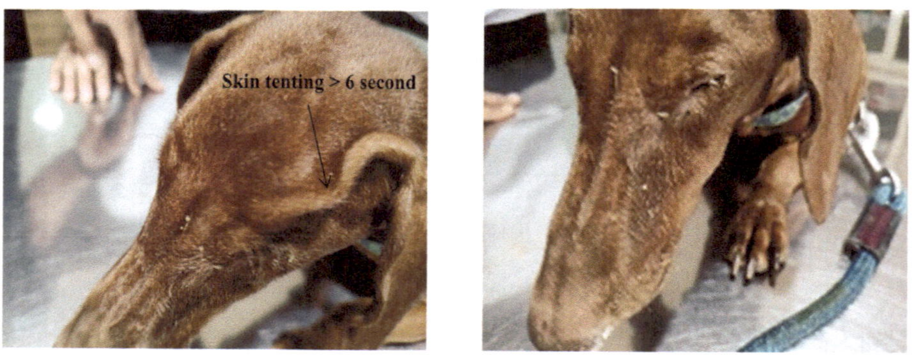

A B

Fig. 28: An adult male Dachshund dog with hypovolemic shock due to severe diarrhoea and vomiting showing prolonged skin tenting (> 6 second –**A**), sunken and closed eyes, reduced mentation, dullness **(B)** ,tachycardia (HR 166 bpm), weak pulse, congested conjunctiva, low rectal temperature (98^0 F), slow respiration, haemoconcentration (Hb, 16.8 g/dl, PCV 55.0%) , neutrophilic leukocytosis (TLC 27,000/mm^3, Neutrophil 90%) suggesting late decompensatory stage of hypovolemic shock.

Cardiogenic Shock

- This type of shock is associated with cardiovascular disease hampering cardiac output as is seen in dogs with heartworm disease, valvular disease, congestive heart failure, dilated or hypertrophic cardiomyopathy, and severe arrhythmias.
- Drugs causing depression of the myocardium may also lead to cardiogenic shock.
- Exercise intolerance, weakness, syncopal episodes, and depression is the usual history in such cases.
- The symptoms of cardiogenic shock follow the typical progression pattern from the compensatory stage to the terminal stage.
- Cardiac murmurs, arrhythmias, jugular distension, dyspnoea, rales and/ or crackles in lungs are common.
- Dogs/cats with cardiogenic shock are very fragile needing no further stress.
- Oxygenation and improving cardiac output are the key to treatment in cardiogenic shock. Antiarrhythmic, diuretics, inotropes are the other medicines used in dogs with cardiac disease. Details are given under heart failure.

Distributive Shock

- It is commonly seen in dogs with anaphylaxis (Fig.29 A, B) and sepsis.
- Its predominant features are vasodilatation and subsequent blood pooling in peripheral veins.
- Release of inflammatory mediators and toxins into circulation increases vascular permeability resulting into loss of fluid into interstitial spaces.
- Cytokine induced vasodilatation of peripheral veins is responsible for shunting of blood from the brain, heart, liver and kidneys to the skin leading to compromised perfusion and oxygenation of vital organs.
- Dogs and cats with distributive shock initially show signs (tachycardia, strong pulse, injected mucus membranes, rapid capillary refill time and hyperthermia) associated with hyperdynamic state as seen in compensatory shock.

- With the progression of distributive shock, decompensatory (early and late) stages become prominent.
- Treatment with aggressive fluid therapy (colloids), oxygen therapy, and positive inotropes (dopamine or dobutamine) needs to be initiated.
- Specific treatment to deal anaphylaxis or sepsis is necessary to deal with the underlined etiology.

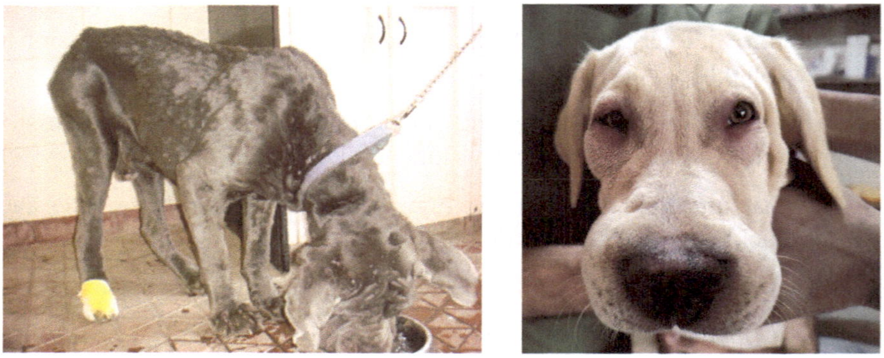

A B

Fig. 29: Mastiff **(A)** and Labrador **(B)** dogs with distributive shock due to anaphylactic reaction showing urticarial plaques all over the body **(A)** and marked swelling on face, lips and eyelids **(B),** hyperthermia (103.5^0 F , 104.^{0}F), tachycardia (145 bpm, 170 bpm), strong pulse, injected conjunctiva and lowered capillary refill time .

Metabolic Shock

- It is seen in patients where oxygen delivery to cell is normal but cell is unable to utilize oxygen for energy production.
- Metabolic shock is seen in cases of hypoglycaemia, cyanide poisoning and mitochondrial dysfunction (Hopper *et al.*,2012).

Hypoxic Shock

- There is impaired oxygen delivery in this type of shock.
- This type of shock is secondary to a decrease in oxygen content of the blood as seen in anemia (decreased hemoglobin concentration), decreased hemoglobin saturation or respiratory disease.
- In some cases oxygen content of the blood may be normal but oxygen offloading may be inadequate.

Diagnosis of Shock

The diagnosis of shock is centered around a careful physical examination, blood pressure measurement and monitoring of blood lactate (a marker of anaerobic respiration).

- Examination of perfusion parameters (heart rate, pulse quality, capillary refill time, and color of the mucus membranes) is of great diagnostic assistance.

General Therapeutic Protocol

The aim of the treatment of shock is to restore perfusion of tissues/organs, improve oxygenation and to address the underlying etiology.

- Oxygen therapy is an important consideration and should be instituted as per requirement. Oxygen can be provided using flow by oxygen (oxygen concentration 25% to 45%), masks (35% to 55% delivered @ 6.0 to10.l/minute), nasal cannulas (30.0 -50.0% @ 100.0-150.0 ml/kg/minute using a single cannula) or an oxygen cage (40.0-50% oxygen). Details are given under chapter Oxygen therapy.
- Intravenous fluid therapy (combining crystalloids and colloids) tailoring to the need of individual dog/cat is an important consideration in the management of shock. Crystalloids can cause fluid overload if used aggressively. High volumes of isotonic crystalloid (0.9% saline) given aggressively may cause hypernatremia and hyperchloremia. During resuscitation, Ringer's lactate is a better crystalloid. Hypertonic saline (7.0%) can expand plasma volume with small volume of solution and it is an ideal choice in the case of head trauma and cardiac disease where large volumes may not be tolerated (Peterson and Moses, 2011).Hypertonic saline can be safely administered @ 2.5 to 5.0 ml/kg IV slowly (1.0 ml/kg/minute). Prolonged use of hydroxyethyl starch (colloid) has shown to increase the risk of acute kidney injury, coagulopathy and mortality in human (Cazzolli and Prittie, 2015). Newer colloids such as tetra starches are considered safe owing to their increased clearance. Blood and blood products are of great value in shocks associated with severe anemia or acute blood loss.
- Cats with shock need to be kept warm through the use of warm water recirculating pads, warm water bottles, warm packs or towels. Overheating should be avoided.

Additional Therapies

- Antibiotics should be administered as soon diagnosis of sepsis is made.
- Gastric protectants viz. proton pump inhibitors pantoprazole, omeprazole), histamine receptor antagonists (ranitidine, famotidine), and sucralfate should be used to prevent or treat stress induced mucosal damage.
- Anti-emetics (ondansetron, maropitant citrate, dolasetron) are required to manage nausea and /or vomiting.

Patient Monitoring

- Shock being a dynamic process, continued monitoring and reassessment of the patient is very much essential.
- Heart rate and rhythm, mentation (mental alertness), pulse rate and its quality, capillary refill time , colour of mucus membranes, and respiratory rate should be closely monitored in a shock patient.
- Estimation of packed cell volume may give an indication about the number of erythrocytes and hemodilution. It can be used to assess fluid resuscitation. Packed cell volume should be maintained between 25 to 35% in critical patients. Fluid resuscitation can also be monitored through the measurement of central venous pressure (CVP) that can be maintained between 8.0 to 10.0 cm H_2O during fluid therapy.
- Pulse oxymetry can be used to monitor oxygen saturation and attempt should be made to maintain oxygen saturation above 94%.
- Urine output is the good indicator to evaluate the efficacy of fluid therapy, determining adequate perfusion and to monitor volume overload.
- Evaluation of total serum proteins or total solids speaks about capillary oncotic pressure (COP). Maintaining a level of total serum proteins > 3.5 g/dl is highly desirable.
- Arterial blood pressure should be frequently monitored.
- Monitoring plasma lactate level is a good indicator of tissue perfusion and oxygenation as its level increases (> 2.0 mEq/L) in lactic acidosis due to anaerobic metabolism.
- Blood glucose, electrolytes and blood gas measurements provide useful information in shock cases.

References

Boulpaep, E.L. (2009). Regulation of arterial pressure and cardiac output. In: Medical Physiology: A Cellular and Molecular Approach. Boron, W.F., and Boulpaep, EE.L. (eds).2nd edn., Philadelphia : Elsevier pp 554-576.

Cazzolli, D. and Prittie, J. (2015).The crystalloid-colloid debate: consequences of resuscitation fluid selection in veterinary critical care. *J. Vet .Emerg. Crit .Care* 2015; 25(1):6-19.

Hopper,K.,Silverstein,D.C., and Bateman,S. (2012). Shock syndrome. In: Fluid , Electrolyte and Acid Base Disorders in Small Animal Practice. DiBartola,.S.P.(ed.) 4th edn. St.Louis:Elsevier 557-583.

Peterson, N.W. and Moses, L. (2011). Oxygen delivery. Compend.Contin.Edu.Pract.Vet. 33: E1-E7

8

Syncope

Syncope can be aptly defined as sudden transient loss of consciousness and postural tone due to an abrupt cerebral hypoperfusion leading to a decreased availability of essential nutrients like glucose. Recovery from syncope is mostly spontaneous without any medical intervention. Actual prevalence rate of syncope in dogs and cats in India is not known. Outside estimates based on referral data base indicate incidence of syncope in dogs and cats as 0.15 and 0.03% respectively. Episodic generalized or hind limb weakness, ataxia or changed consciousness is termed as pre-syncope. Decreased or short breaking off blood to the brain results into Syncope. Severe arrhythmias and conduction disturbances are in the core of the problem in syncopic dogs and cats.

Etiology

- Etiology of the syncope in dogs and cats can broadly be grouped in to three categories such as neurally mediated, cardiac mediated, and metabolic disorder mediated (Fig. 30).

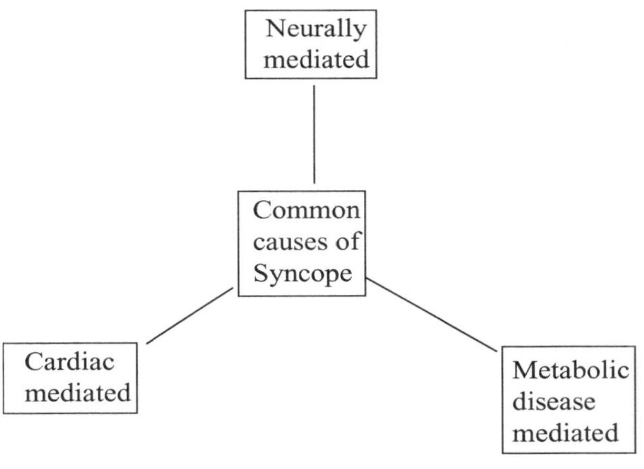

Fig. 30: Diagrammatic representation of common causes of syncope.

- Neurally mediated syncope is circumstantial and may be observed during coughing, defecation, urination, gagging, eating too fast, barking, extreme excitement/exertion, and vomiting. Severe anemia can lead to exertional syncope or pre syncope. Coughing syncope occurs most often in brachiocephalic dogs or dogs with airways disease. Neurocardiogenic (vaso-vagal) syncope are less common in animals and are secondary to sudden bradycardia following a bout of tachycardia in small breeds of dogs. Vaso-vagal, emotional, carotid sinus and situational syndrome falls under the category of neurally mediated syncope. Recently Fukushima *et al.* (2018) have reported a case of syncope associated with swallowing in an intact male West Highland White Terrier dog that responded after pacemaker implantation.

Cerebral edema, brain tumour, meningitis/encephalitis, and cerebral vascular obstructions are the potential cause of increased intracranial pressure that can lead to decreased cerebral perfusion resulting into neurologic syncope. A case of syncope with bradycardia due to a large meningioma has been described in a four year old Irish setter dog (Freeman *et al.*,1985).

- Cardiac mediated syncope is generally observed during severe cardiac arrhythmias and conduction disturbances (bradycardia, hyperkalemia induced bradycardia, tachycardia, sick sinus syndrome, atrial fibrillation, atrial standstill, supraventricular tachycardia, ventricular fibrillation/ ventricular tachycardia, advanced 3rd degree heart block, dysfunctioning of implanted pace maker), cardiac diseases (cardiomyopathies, pericardial diseases, atrial filling defect, tamponade), obstruction to left ventricular outflow (aortic stenosis, hypertrophic cardiomyopathy, intracardial tumor), and obstruction to right ventricular outflow (pulmonic stenosis, pulmonary hypertension, heart worm disease, pulmonary thromboembolism cardiac tumors). Cardiogenic syncope are of life threatening nature.
- Metabolic disease/disorder mediated syncope can be observed in dogs and cats with severe hypoglycemia, Addison's disease, acute blood loss (hemorrhage), anemia, acute hypoxia and intoxication with certain drugs/toxins. Hypoglycemic syncope episodes in animals are often caused by the presence of insulinoma. Metabolic mediated syncope in cats is uncommon and occasionally seen in diabetic cats owing to hypoglycemia due to higher dose of antidiabetic drugs. Hypocalcaemia and hypernatremia are the other causes of syncope.
- In cats syncope is usually related with arrhythmias (high grade 2nd or 3rd degree heart block).Syncope due to metabolic causes in cats is uncommon.

Predisposing factors

- Syncope are more common in old dogs.
- Sick sinus syndrome in cocker spaniels, miniature schnauzers, pugs and dachshunds; and ventricular arrhythmia in boxers and German shepherds may predispose them to syncope.
- Cardiogenic syncope is more common in animals.
- Most non-cardiogenic syncope are not life threatening.
- Cardiogenic syncope have higher risk of mortality.

Pathophysiology

- The nature of the syncope is sudden and acute.
- Cardiogenic syncope is related to reduce cardiac output owing to arrhythmias, decreased cardiac filling, or obstruction of blood flow from the heart.
- Hypoxia, hypoglycemia with normal cerebral blood flow, or decreased vascular resistance is responsible for neurocardiogenic syncope.
- Majority of cases in dogs and cats are related to sudden transient decrease in cardiac output leading to reduced blood flow in the brain.
- Pathophysiology of situational syncope involves accelerated vagal nerve activity and bradycardia, reduced blood pressure owing to decline in sympathetic nerve activity and reduction in cardiac preload (Benditt, 1997 and Moya *et al.*,2009)

Symptoms

Dogs and cats with syncope predominantly show following signs in different permutation and combination

- Sudden fainting or collapse
- Sudden loss of consciousness
- No responsiveness
- Stiff limbs
- Generalized or hind limb weakness
- Posterior ataxia

- Muscle spasms may or may not be seen
- Pale mucus membranes
- Usually no defecation or urination during syncope episode
- Spontaneous sudden recovery

Diagnosis

- Though syncope is a sudden transient and temporary event, diagnosing underlying etiology is very much important. History and clinical examination may provide clue for determining underlying etiology. History, clinical examination, situational activity and symptoms occurring during and after event play a crucial role in ascertaining etiology of the syncope. Episodic weakness can be differentiated from syncope as there is no loss of consciousness in the episodic weakness. Episodic weakness can also occur with cardiac diseases or hypoglycemia.
- Detailed clinical examination, electrocardiographic and echocardiographic examinations, routine laboratory tests (CBC, blood smear examination, detailed serum biochemistry and urological examinations) may be of diagnostic value in revealing underlying etiology.
- Blood pressure should be monitored in cases of syncope to rule out hypotension specially when the dog is on beta blockers. A normal blood pressure does not rule out transient hypotension that occurs only during syncope episode.
- Electrocardiography can be employed for detecting arrhythmias. Sometimes arrhythmias and syncope are so infrequent that a long 10-20 minute electrocardiogram may be required. Exercise induced coughing or vagal maneuvers should be done before or during electrocardiography in suspected cases of cardiogenic syncope.
- Echocardiography is a better choice if structural or functional cardiac disorders are suspected as a cause of syncope.
- Thoracic radiography needs to be performed for evaluating lungs, pleural space, mediastinum, pulmonary vessels, cardiac size and signs of pulmonary disease.
- Bleeding hepatic and splenic masses can be detected with abdominal sonography.

- Level of cTn-I in an individual is not optimally discriminatory to differentiate cardiogenic syncope from collapse with generalized epileptic seizures or vasovagal syncope (Dutton *et al.*, 2017).

Prognosis

- Prognosis of the cases of syncope is dependent on the underlying cause.
- Mortality in cases with cardiogenic syncope is higher.
- Syncopic animals with structural cardiac disease have poor prognosis.
- Syncopic Boxers and Doberman with cardiac disease are at an increased risk of sudden death.
- Prognosis of cases with neurocardiogenic syncope is unknown as detail studies are lacking.
- Sudden death in animals with cough syncope is rare.

Treatment

- Treatment is directed at managing the underline cause.
- Syncope due to arrhythmias is to be treated with antiarrhythmic drugs.
- Syncope due to ventricular tachyarrhythmia can be managed with lidocaine or sotalol.
- Syncope due to severe bradyarrhythmias may be attended with atropine, glycopyrrolate or pace maker implant.
- Dog and cats having cardiogenic syncope are required to be monitored periodically and continuously using electrocardiography.
- Beta blockers have shown promise in humans with neurocardiogenic syncope. But they may precipitate bradycardia related syncope in Doberman dogs with cardiomyopathy.
- Neurally mediated syncope with pulmonary hypertension or high preload may respond favorably to cardiac offloading therapy.
- Exertional/ excitement mediated syncope with murmurs in small breed dogs may be due to impending congestive heart failure and may require treatment with diuretics, phosphodiesterase inhibitor III (PDE-III inhibitor) and ACE- inhibitors.
- Avoiding exertion/ excitement is a key to mange excitement mediated syncope.

References

Benditt D. G.(1997). Neurally mediated syncopal syndromes: pathophysiological concepts and clinical evaluation. *Pacing* Clin. Electrophysiol. 20: 572–584. doi: 10.1111/j.1540-8159.1997.tb06211.x [PubMed] [Google Scholar]

Dutton,E.,Dukes-McEwan,J. and Cripps,P.J. (2017). Serum cardiac troponin-I in canine syncope and seizures. J. Vet.Cardiol.19:1-13. Doi: 10.1016/j.jvc. 2016.10.001. Epub 2016 Dec.5.

Freeman,K.P., Monlux, A.W.,Herald, D. and Sealock, M.C. (1985). Bradycardia associated with meningioma in a dog. J.Am.Vet.Med.Assoc.187:838-839.

9

Drugs in the Management of Heart Failure

Pathogenesis of heart failure in animals and humans is a complex phenomenon. The major abnormality in heart failure is myocardial contractility failure, disturbances in rhythm and rate of the heart, disturbances of pre load and after load, forward failure, backward failure, fluid accumulation, RAAS activation, hypoperfusion of organs, oxidative stress, activation of proinflammatory cytokines etc. Therefore, drugs of different groups are required to address these problems. For the convenience details of some of the important and common drugs, used in the management of cardiac diseases and heart failure, are given below.

ACE-Inhibitors

ACE-inhibitors are widely used in the treatment of chronic congestive heart failure in dogs and cats. These drugs prevent retention of fluid and sodium; prevent vasoconstriction and myocardial fibrosis; and are mildly hypotensive. These are balanced vasodilators and reduce both pre load and after load. ACE-inhibitors are contraindicated in pre-existing hypotension, hypovolemia, acute renal failure, valvular stenosis, and cardiac tamponade. In dehydrated patients ACE-inhibitors can cause acute renal failure. Side effects of ACE-inhibitors are hyperkalemia and azotemia. ACE –inhibitors can cause cough in humans but it is not common in animals. Enalapril, benazepril, lisonopril and ramipril are the commonly used drugs in this group.

Benazepril

Benazepril, is a prodrug that is converted to active metabolite (benazeprilat) in the liver. It is used in the management of congestive heart failure, hypertension, chronic renal failure and protein losing nephropathy. It has shown promising results in improving exercise tolerance, improving quality of life and prolonging life in humans with heart failure. Benazepril is metabolized by the liver into its active form benazeprilat that blocks the conversion of angiotensin I (AT-I) to angiotensin II (AT-II). After oral administration benazepril is rapidly absorbed and eliminated through kidneys. Although the drug is metabolized by the liver and converted into active metabolite benazeprilat, compromised liver does not appear to significantly reduce the level of benazeprilat. Benazeprilat is excreted almost equally in the bile and urine. Therefore in renal compromised patients with congestive heart failure, benazepril is more appropriate drug than the enalaril. Benazepril alone can be used to manage hypertension in cats. A combined use of benazepril and amlodipine is more effective in the management of severe hypertension in cats. A combination of benazepril with spironolactone has recently been approved by USFDA (USA) for the use in the management of heart failure associated with myxomatous mitral valve disease in dogs. Benazepril is also a drug of choice in the management of chronic renal failure and proteinuria in dogs and cats. Side effects of benazepril are hypotension, weakness, lethargy, gastrointestinal signs (anorexia, diarrhea, vomiting), hyperkalemia and renal dysfunction. Though it is indicated in the management of chronic renal failure, in some instances mild to moderate reversible renal failure can be precipitated owing to volume depletion. Therefore patients receiving benazepril are required regular monitoring for blood pressure, serum electrolytes and renal function. ACE-inhibitors are better avoided in acute renal failure, pre-existing hypotension, hypovolemia, or hyponatremia. Benazepril should not be used in lactating or pregnant animals. ACE-inhibitors increase the effect of diuretics. Potassium sparing diuretics or potassium supplementation in conjunction with benazepril may increase the risk of hyperkalemia. Animals receiving benazepril should not be given NSAIDs (including aspirin) due to increased risk of acute renal failure. Concurrent use of benazepril and insulin may increase the risk of hypoglycemia.

Captopril

It is also an ACE-inhibitor that prevents conversion of angiotensin-I (AT-I) to angiotensin –II (AT-II), It reduces circulating plasma aldosterone concentration and acts as an arteriolar and venous dilator. It is not a good drug to be used in emergency in dogs with severe pulmonary edema as it reduces the diuretic and natriuretic effects of furosemide. In humans it has been used to treat all stages of heart failure. In dogs it has been used in the management of acquired mitral regurgitation. It can be given @ 0.5 to 1.0 mg/kg PO TID. The drug is well tolerated. Its side effects are anorexia, vomiting, diarrhea, hypotension and azotemia.

Enalapril

Enalapril is another ACE- inhibitor that blocks conversion of angiotensin I (AT-I) to angiotensin II (AT-II). It is commonly prescribed in the management of hypertension and heart failure in dogs and cats. It is also being used in the management of chronic renal failure and protein losing nephropathy. It also reduces fluid build-up in the lungs. After absorption from gastrointestinal tract enalapril is converted to active metabolite enalaprilat in the liver. Enalapril and its metabolite enalaprilat both are primarily excreted through kidneys. Therefore enalapril is not a drug of choice in dogs and cats with cardio-renal failure. In treating congestive heart failure, enalapril is usually combined with furosemide, spironolactone, or positive inotropes. It can also be combined with beta blockers in the management of occult dilated cardiomyopathy or systemic hypertension. Though, amlodipine is the drug of choice in cases of severe hypertension in cats, enalapril can be used in cases of mild hypertension. Adverse reactions of enalapril may vary from mild to severe. Diarrhoea, cough, decrease in appetite, vomiting, dizziness, drowsiness, rashes, dry mouth ,sleeping disturbances, frequent urination, fainting, fever, chills,, bleeding, weight gain, bruising, and swelling are the adverse effects of enalapril. When enalapril is used with other diuretics and vasodilators, chances of hypotension and hyperkalemia increases. Therefore dogs and cats receiving enalapril require regular monitoring as described under benazepril. Enalapril should be avoided during pregnancy owing to birth defects. Use of aspirin is contraindicated during enalapril therapy. Anesthetics, antacids, antihypertensive drugs, vasodilators, baclofen, buspirone, cabergolline, cimetidine, corticosteroids, erythropoietin, digoxin, diphenhydramine, diuretics, heparin, NSAIDs, opioids, potassium, potassium sparing diuretics, probenecid, sildenafil are to be used cautiously with enalapril. Patients on regular enalapril should be regularly monitored for blood pressure, kidney function, electrolytes and urine protein.

Lisonopril

Lisonopril is also an ACE- inhibitor having vasodilation activity. It is a prescription drug and is used in the management of high blood pressure or heart failure like other ACE-inhibitors. Though, it is not approved by USFDA, it is being used as off level drug at the dose rate of 0.5 mg/kg PO orally once daily in dogs. In cats the dose of lisonopril varies from 0.25 to 0.5 mg/kg orally once daily. Its adverse effect include vomiting and diarrhea. Its use is not common in veterinary practice.

Ramipril

Ramipril is another ACE-inhibitor used in the management of congestive heart failure grade II, III, and IV caused by valvular insufficiency due to chronic degenerative valve disease or cardiomyopathy with or without adjunct therapy with furosemide. It is given @ 0.125 mg/kg once daily in dogs. It can be given with food. It is contraindicated in dogs with aortic stenosis or mitral valve stenosis. The drug should be discontinued if signs of apathy or ataxia appears during treatment and once symptoms disappear drug administration can be resumed at 50% level of the original dose. It has been shown that the addition of ramipril to pimobendan and furosemide did not have any beneficial effect on survival time of dogs with congestive heart failure due to MMVD (Wess *et al*., 2020).

Anti-arrhythmic Drugs

A. **Anticholinergics-** This group of drug includes atropine, glycopyrrolate and propantheline bromide and is used to increase SA node activity and AV conduction when bradyarrhythmias are due to excessive vagal tone.

Atropine

Atropine is an alkaloid extracted from belladonna plant. It belongs to anticholinergic and antispasmodic group of drugs. It is a prescription drug commonly used in the management of bradycardia, increased salivation, increased bronchial secretions, or as an antidote for the overdose of cholinergic drugs, mushroom poisoning, toxicity of organophosphorus and other parasympatheticomimetic compounds. An atropine challenge test is done to confirm the degree of vagal influence on SA node and or AV node function. Parenteral administration of atropine (0.02 mg/kg IV) can temporarily exacerbate vagally mediated AV block. In this situation atrial rate increases fast.

Glycopyrrolate

Glycopyrrolate is indicated in the management of ventricular escape rhythm, sinus bradycardia, second degree heart block (AV block) or SA arrest @ 0.005 to 0,01 mg/kg IV or IM. Unlike atropine, it does not have centrally mediated effects. Effects of glycopyrrolate last longer than atropine.

Propantheline Bromide

Propantheline bromide is an oral anticholinergic. In dogs it is used @ 3.75 to 7.5 mg PO BID/TID.

B. **Sympathomimetics-** Terbutaline and isoproterenol are the drugs in this group which have been used in the management of symptomatic AV blocks or atropine unresponsive bradycardia.

Terbutaline

Terbutaline is a synthetic sympathomimetic amine derived from resorcinol. It is beta agonist with broncho dilating properties. It is mainly used in the management of airways disease (asthma, bronchitis) in dogs and cats. It can weakly stimulate SA node and beta receptors in the heart. Because of its stimulating effect on SA node, it has been used in the management of atrial standstill, sick sinus syndrome or AV block where pace maker therapy is not feasible. Kumar and Varshney (2005) have reported beneficial effect of terbutaline (0.1 mg/kg PO BID)in a German Shepherd dog with atrial standstill. Beta blockers may antagonize the effect of terbutaline. When terbutaline is used with digoxin, it may increase the risk of arrhythmias. Side effects of terbutaline are dose dependent and include gastrointestinal disturbances (anorexia, vomiting, diarrhea), restlessness, tremors, excitement, and increase in heart rate. Terbutaline should be used with caution in cases having history of heart failure, diabetes mellitus, seizures, hypertension , hyperthyroidism or those on thyroxin therapy.

Isoproterenol

Isoproterenol is a beta agonist advocated in the management of symptomatic AV-block or atropine unresponsive bradycardia. Since it can cause hypotension, it is not recommended in the treatment of heart failure or cardiac arrest. Adverse effect of isoproterenol is arrhythmias like other catecholamines. Its dose rate in dogs and cats is 0.045 to 0.09 microgram/kg /min CRI.

C. **Calcium Channel Blockers-**Calcium channel blockers are widely used in the management of cardiac arrhythmias, systemic hypertension, and hypertrophic cardiomyopathy in humans and companion animals. Calcium channel blockers exert their effect on voltage operated L-type channel within the plasma membrane leading to negative inotropic and chronotropic effects in cardiac tissues. Calcium channel blockers were originally used in managing atrial fibrillation in dogs. Now they are becoming drug of choice in the treatment of hypertrophic cardiomyopathy in cats and in the treatment of systemic hypertension. Verapamil, diltiazem, nifedipine and amlodipine are common calcium channel blockers being used as cardiac medicine in companion animals.

Verapamil

Verapamil is calcium channel blocker being used in the treatment of supraventricular arrhythmias. It is prototypical phenylalkylamine. It slows down the activity of SA node pace maker, AV conduction and prolongs AV node refractory period (Opie *et al.* , 1995). It is effective in the management of supraventricular tachycardia, atrial flutter, and atrial fibrillation. It is also a potent arteriolar vasodilator and has been used in the management of angina and systemic hypertension in human.

Its dose rate in dogs is 0.5-1.0 mg/kg PO TID. For supraventricular tachycardia it may be given intravenously slowly @ 0.05 mg/kg with monitoring of blood pressure and electrocardiogram. Adverse effect of intravenous therapy are AV-block and hypotension. Verapamil is contraindicated in sick- sinus syndrome, pre-existing AV- block, Wolf-Parkinson-White syndrome, myocardial failure, and digitalis toxicity. Verapamil can potentiate negative inotropic effect of beta adrenergic blockers and quinidine.

Diltiazem

Diltiazem, a benzothiazepine derivative, is a calcium channel blocker used in the management of arrhythmia such as supraventricular arrhythmias. It slows down the activity of SA node pacemaker, and conduction through AV node. It also causes frequency dependent channel blockade in AV- node. It also improves coronary blood flow. It is also recommended in the management of hypertrophic cardiomyopathy in cats. In canine atrial fibrillation diltiazem can be used along with digoxin. Atrial fibrillation is common in dogs with DCM or DMVD. Though adverse effects are not common at standard dose rate, decreased appetite and slow heart rate are sometimes observed with diltiazem therapy. Diltiazem can be given at the dose rate of 0.5 to 1.5 mg/kg PO TID in dogs and cats. For the treatment of supraventricular tachycardia (SVT), it can be given intravenously (0.25 mg/kg over 2 minutes). When it is used with cimetidine, bioavailability of diltiazem is increased and clearance is decreased. Concurrent use of propranolol reduces clearance of diltiazem. Though adverse effects of diltiazem are rare, bradycardia, hypotension and gastrointestinal upset may be observed.

Nifedipine

Nifedipine, 1st generation dihydropyridine, has arteriolar vasodilatory effect. It has no effect on depolarization of AV node or AV node conduction. It is not effective in the treatment of supraventricular tachycardia. It is contraindicated in cases of hypertrophic obstructive cardiomyopathy as peripheral vasodilation may exacerbate the existing outflow obstruction. Because of its profound vasodilation and consequent hypotension it is not being used frequently in veterinary practice.

Amlodipine

Amlodipine, 2nd generation dihydropyridine, is slow in action and has long lasting effect as compared to nifedipine. It has little effect on myocardial contractility or cardiac conduction in therapeutic doses. It is effective in reducing systolic and diastolic blood pressure and systemic vascular resistance without inducing reflex tachycardia. In renal hypertensive dogs it has been used @ 1.0 mg/kg as a single dose to reduce blood pressure. In cats it is used @ 0.625 -1.25 mg/cat once daily to manage systemic hypertension. Side effects of amlodipine are not known in cats. However ankle edema is common side effect in humans.

D. **Beta-Blockers-**These drugs antagonize the effect of sympathetic stimulation thereby slows heart rate and reduces myocardial oxygen demand. These are class II antiarrhythmic drugs. In humans, beta blockade is recommended virtually in all instances of reduced contractility, such as occurs in DCM. Metoprolol and carvedilol have been used in the management of dilated cardiomyopathy in humans. Because of this information, beta blockers are being attempted in the management of canine dilated cardiomyopathy. Because of the risk of acute slowing of heart rate, and decrease in myocardial contractility , use of beta blockers in canine DCM does not seem logical. Use of beta blockers in DMVD is also controversial. Beta blockers can be used in conjunction with digoxin in the management of atrial fibrillation/atrial flutter. Beta blockers are also drugs of choice in the management of hypertrophic cardiomyopathy in the cats. Initially low doses of beta blockers are advised. Then the dose may gradually be increased to the effect. Adverse effects of beta blockers include slowing of heart rate, low blood pressure, worsening of heart failure, depressed attitude, bronchospasm (more likely with propranolol) and masking early signs of hypoglycemia in diabetic animals. Commonly used drugs in this groups are atenolol, propranolol, sotalol, carvedilol.

Atenolol

It is selective beta-1 blocker. It has been used in the management of clinical heart disease to slow sinus rate and AV conduction and to suppress premature ventricular depolarization. It is excreted in urine. In renal compromised animals its clearance is delayed. It does not cross blood brain barrier. Its adverse effects includes weakness or exacerbation of heart failure.

Propranolol

It is non-selective beta blocker being used in dogs and cats. It is used to reduce arterial blood pressure. It should be used cautiously in animals with heart failure. In cases of myocardial failure prior digitalization is recommended. It has potential brochoconstrictive property. Therefore its use in congestive heart failure should be delayed until pulmonary edema is resolved. Bradycardia, hypotension,, heart failure, bronchospasm and hypoglycemia are some of its toxic effect.

Sotalol

It is a non-selective beta blocker and class-III anti-arrhythmic drug. It is used in the management of supraventricular and ventricular tachycardia and Boxer arrhythmogenic cardiomyopathy @ 1.0-2.0 mg/kg PO BID. Sotalol is generally used as monotherapy but can be combined with mexiletine if rhythm control is inadequate with sotalol alone. Though it has some effect in atrial fibrillation but not very efficacious. It can safely be used in combination with ACE-inhibitors, furosemide, spironolactone and pimobendan. It should not be combined with other beta blockers (atenolol), other negative inotropes (diltiazem), or other class III antiarrhythmic drugs (amiodarone). Its adverse effects are negative inotropy, bradyarrhythmias (sinus and AV block) and proarrhythmia.

Carvedilol

Carvedilol is a non-selective alpha and beta blocker having oxygen scavenging capability and ability to reduce endothelin release. Carvedilol not only negates the effects of sympathetic nervous system (SNS), it also reduces heart rate, causes vasodilation and acts as an antioxidant. The drug is a negative inotrope and expensive. The drug can be started early in the disease process with low doses (1.56 mg PO OD) titrating upward towards a total target dose of 25 mg PO BID in large breeds of dogs. If dog cannot tolerate increases in carvedilol dosage, last tolerated dose can be accepted as maximum dose. Currently carvedilol is recommended @0.25 mg/kg BID increasing every two weeks until a target dose of 0.8-1.0 mg /kg BID is achieved. Its use should be avoided in phase IV heart Failure (NYHA).

E. **Other Drugs**

Amiodarone

It is antiarrhythmic drug used to convert atrial fibrillation back to normal rhythm. It is categorized as class –III antiarrhythmic drug. Its side effects are loss of appetite, vomiting, bradycardia, induction of liver enzymes and destruction of platelets. Other short and long term adverse effects of amidarone are pulmonary fibrosis, bradycardia, thyroid dysfunction, ocular abnormalities, peripheral neuropathy and gastrointestinal upset. It has been used in dog in loading and maintenance dose of 16.5 and 9.0 mg/kg PO OD respectively

Lidocaine (Xylocaine)

It is class IB antiarrhythmic drug that is used predominantly in the management of ventricular arrhythmias (ventricular tachycardia, ventricular premature complexes). It has no effect on supraventricular arrhythmias. Because of short half life it is ideal for the treatment of acute ventricular arrhythmias. It is metabolized by the liver. Therefore compromised liver can prolong its half life. It is given @ 2.0 mg/kg IV as a bolus over one minute to effect or to a cumulative dose of 8.0 mg/kg over 30 minutes. To maintain rhythm control, boluses are to be repeated or the drug is to be given as CRI (25-75 microgram/kg/min). It is rarely indicated in cats (0.1 to 0.4 mg/kg IV as bolus, then 10-20 microgram/kg/min as CRI). Drowsiness, agitation, muscle twitching, convulsions, hypotension are some of its adverse effects. Cats are more susceptible to its toxicity. Cardiac suppression and CNS excitation may be seen.

Mexiletine

It is an oral analogue of lidocaine. It is recommended in the treatment of ventricular arrhythmias in dogs and is a drug of choice in dogs who do not tolerate sotalol. It can also be used as an adjunct in severe ventricular arrhythmias not controlled by sotalol alone. The dose rate of mexiletine is 4.0 to 6.0 mg/kg PO TID. It should be given with food to limit toxicity. Anorexia, vomiting, tremors, and hepatic toxicity are its some adverse effects. Liver function should be evaluated periodically when the drug is used for long term.

Tocainide

It is also an analogue of lidocaine and recommended for similar indications as lidocaine. It is rarely advised as monotherapy for ventricular arrhythmias. It is mostly used as an add on therapy to sotalol. Because of high cost the drug is not in common use. Potential adverse effects are CNS and GI disturbances, hypotension, arrhythmias, and progressive corneal edema. Its dose rate in dogs is 15–20 mg/kg, PO TID.

Phenytoin

It is primarily indicated in the management of digitalis-induced arrhythmias, because it abolishes digitalis-induced abnormal automaticity. The recommended dosage in dogs is 30 mg/kg, PO, TID, or 10 mg/kg IV over 5 min.

F. Diuretics

Diuretics are the cornerstone of heart failure therapy. Extensive diuresis also activates RAAS system. Diuresis as a monotherapy is not advisable and doses of diuretics should be kept minimum to avoid RAAS activation, dehydration, azotemia and hypokalemia. When diuretics are used with ACE-inhibitors, the dose of diuretics can be reduced to half. Diuretics are advocated in the late phase II and III heart failure.

Furosemide

Furosemide is a loop diuretic and is the drug of choice for pulmonary edema (life threatening or congestion) in congestive heart failure. Because it activates the RAAS, lowers blood pressure and cardiac output, causes azotemia and electrolyte disturbances, and potentiates adverse effects of other cardiac therapies, it should not be used as monotherapy but should be used in conjunction with ACE-inhibitors. Development of azotemia in dogs and cats receiving furosemide along with other drugs for the management of congestive heart failure requires either lower doses of furosemide or complete cessation of furosemide. It is the drug of choice in the emergency management of pulmonary edema. Diuresis starts within 5 minutes of intravenous administration of furosemide. After oral dosing diuresis begin within 1 hour. In refractory heart failure and heart failure with hypokalemia, use of potassium sparing diuretics is recommended. Addition of torsemide with reduction in dose of furosemide may be beneficial in such cases. Common adverse effects of furosemide are losses of excessive fluid and electrolytes resulting into dehydration and marked weakness.

Spironolactone

Spironolactone is an aldosterone antagonist and is receiving renewed interest because of its promising results in the management human heart failure. Because spironolactone is a weak diuretic, its benefits are possibly due to blunting the adverse effects of aldosterone. It may reasonably be used in an early stage of heart failure. But data on its use in an early stage of heart failure or pre-heart failure states is lacking. Elevated aldosterone level has been observed despite the use of ACE- inhibitors in both humans and dogs with heart failure. It seems logical that suppression of elevated aldosterone by spironolactone can be beneficial in dogs with DMVD. No increase in electrolyte abnormalities has been reported with concurrent ACE-inhibition and aldosterone blockade. Therefore spironolactone can compensate for temporary or incomplete suppression of aldosterone secretion by ACE-I and should be used concurrently with ACE-I. Spironolactone can also be used as an adjunctive diuretic at 1-2 mg/kg QD-BID with a loop diuretic, such as furosemide. Cardiologists in USA prefer lower doses of spironolactone (0.5-1.0 mg/kg PO OD) or coupling of spironolactone with hydrochlorothiazide (thiazide group diuretic). Enhanced diuresis can be evident when two diuretics work synergistically in different parts of the nephron. Adverse effects of spironolactone are potassium retention and gastrointestinal tract upset. When it is used with ACE-inhibitors or oral potassium supplement, regular monitoring of blood potassium level is advisable.

Triamterene

Triamterene is another potassium sparing diuretic that acts on distal tubule. It reduces the reabsorption of sodium and decrease excretion of potassium and hydrogen ion. It has limited action on aldosterone. It is used @ 2.0-4.0 mg/kg per day.

Thaizides

Thiazide group of diuretics (chlorothiazide, hydrochlorothiazide) increases calcium absorption with decrease absorption of sodium and chloride. Diuretics of this group induce mild to moderate diuresis. These can be used in the early stage of heart failure. As thiazide group of diuretics decrease renal blood flow, they should not be prescribed in dogs having pre-existing azotemia. In non-azotemic patients adverse effects of thiazide diuretics are uncommon. If the animal is anorectic or thiazide diuretics are used in excess, hypokalemia, other ion disturbance or dehydration may occur. Hyperglycemia may occur in diabetic or prediabetic animals.

Torsemide is a novel and more potent diuretic than furosemide. It promotes renal excretion of sodium, chloride and water. Greater bioavailability of torsemide can result in stronger, more sustained and uniform diuresis. Torsemide can be used in the management of advanced congestive heart failure. It also has anti-aldosterone effect similar to spironolactone. Careful monitoring of renal function and electrolyte balance is needed as some cases may require adjustment in doses of torsemide..

G. Inotropes

Positive inotropes are the drugs of choice for increasing the contractility of the myocardium. In this group digoxin is the oldest positive inotrope and has been in use in the management of heart disease for long time. Dobutamine, dopamine and amrinone are the other drugs used intravenously as a constant rate infusion. Recently pimobendan is enjoying reputation of a novel drug in the management of congestive heart failure.

Digoxin

Digoxin is a weak positive inotropic drug that mildly increases myocardial contractions. It is the digitalis glycoside and is the drug of choice in arrhythmias such as supraventricular premature complex, supraventricular tachycardia, and atrial fibrillation/flutters. It is also indicated in heart failure caused by myocardial dysfunction (poor or reduced myocardial contractility), chronic mitral valve insufficiency and other chronic volume or pressure overloads. It is contraindicated in cases of hypertrophic cardiomyopathy especially those with ventricular outflow obstruction, serious ventricular arrhythmias, disease of AV node. Digoxin is not useful in pericardial disease. Absorption of digoxin is decreased if it is given with food, antacids or kaolin-pectin. Concurrent administration of quinidine, verapamil and amiodarone increases serum concentration of digoxin. It is therefore advisable to administer digoxin empty stomach. Hypokalemia and thyroid disturbances can potentiate digoxin toxicity. The toxic effects of digoxin can be serious and even life threatening warranting careful dosing. Weekly monitoring of the digoxin level in the blood is recommended to avoid digoxin toxicity. Compromised kidney function, dehydration, loss of lean muscle mass, low blood potassium levels, and certain drugs increase the risk of digoxin toxicity. Heart rhythm disturbance, anorexia, vomiting, diarrhea, and depression are some of the adverse effect of digoxin therapy.

Dopamine

Dopamine exerts marked positive inotropic effects in dogs without affecting the heart rate. Experimental studies in guinea pigs have shown positive inotropic effect of dopamine in normal as well as experimentally induced heart failure. Dopamine is an adrenergic neurotransmitter with specific receptors in the renal vasculature. It has a very short half life and is administered as a constant rate infusion of 2-5 microgram/kg/min. Dilution of 40 mg of dopamine in 500 ml of fluid (saline solution, 5% dextrose in water, or lactated Ringer's solution) yields a solution of 80 microgram/ml. A volume of 0.75 ml/kg/hr will provide 1 microgram /kg/min. Higher dose of dopamine can cause tachycardia, arrhythmia and peripheral vasoconstriction. At low infusion rate (< 3 microgram/kg/min) it produces vasodilation of renal and coronary arteries. In veterinary medicine dopamine is indicated in the treatment of cardiovascular and renal disorders (Goldberg, 1972, Goldberg and Raifer, 1985, Lewis and Bruce, 1988). Low doses of dopamine are recommended for the treatment of oliguria to improve urine output. While high doses of dopamine are recommended to improve hemodynamics in circulatory shock (Giselle,1990).

Dobutamine

Dobutamine is a synthetic analog of dopamine. It is used as a short term inotropic support. It stimulates beta-1 receptors. It increases myocardial contractility with minimum effect on heart rate and blood pressure at lower infusion rate (3 to 7 microgram/kg/min.).

Epinephrine

Epinephrine increases blood pressure, heart rate and myocardial contractility. It can induce arrhythmias. It is the drug of choice for resuscitation after cardiac arrest.

Isoproterenol

Isoproterenol increases myocardial contractility, heart rate and cardiac output by stimulating beta receptors. It decreases blood pressure. It is not to be used in animals with heart failure and cardiac arrest. It can be used in cases of AV block.

Pimobendan

Pimobendan, a positive inotrope, has recently been advocated to increase contractile force of heart muscles by increasing the quantity of intracellular calcium available for binding by muscle proteins or by increasing the sensitivity of contractile proteins to calcium. Pimobendan, a benzimidazole-pyridazinone drug, has been introduced in 2000 and is classified as a balanced inodilator because of its nonsympathomimetic, nonglycoside positive inotropic (through myocardial calcium sensitization) and vasodilator properties (Fuentese, 2004 ; Lefbon 2005;). Its activity is through Troponin-C calcium sensitization. It also causes peripheral arterial dilation, coronary artery dilation, pulmonary artery dilation, and peripheral vanodilation by inhibiting phosphodiesterases III and V in vascular smooth muscles (Hagemeijer, 1993; Chu *et al.,* 1995). As such, pimobendan increases ventricular contractility and reduces pre-load and after-load in dogs with advanced cardiac insufficiency. It increases contractility of cardiac muscles through a mechanism different from that of traditional inotropes such as digoxin. Increase in myocardial contractility without significant increases in myocardial oxygen demand is the major advantage of pimobendan. It is well tolerated in dogs and no specific monitoring is required as in case of digoxin. In failing heart it exert positive inotropic effect primarily through sensitization of cardiac contractile apparatus to intracellular calcium. As a PDE- 3 inhibitor, Pimobendan can potentially increase intracellular calcium concentration and myocardial oxygen consumption. However, the cardiac PDE-3 effects of Pimobendan are minimal in pharmacological doses in dogs with heart diseases, which is the major advantage relative to other inotropic PDE inhibitors. Pimobendan may have some endothelial derived vasodilatory effects and has also demonstrated beneficial modulation of cytokines (tumor necrosis factor –beta, and interleukin 1-beta and 6). Platelet inhibitory effect of pimobendan has also been demonstrated in dogs and cats. PDE3 inhibition in cardiocytes leads to more rapid relaxation, pimobendan can also be considered as a positive luciotrope (drugs having potential to improve ventricular relaxation). Thus pimobendan has become a drug of first choice in positive inotrope group for the management of congestive heart failure in dogs. It is now approved in the United States also for use in dogs with congestive heart failure. The drug is used @ 0.25 mg/kg PO BID. The drug should be administered on an empty stomach for maximum effects when starting therapy. The beneficial effects of pimobendan are due to preload reduction due to vasodilation, afterload reduction by balanced arteriovenous dilation, enhancing contractility without significant increase in oxygen demand, reduction in pulmonary hypertension, modulation of cytokines to improve survivality, its safety with concurrent use of other drugs, and excellent safety profile. Pimobendan has shown beneficial effects in the treatment of canine congestive heart failure, pre-clinical dilated cardiomyopathy, and preclinical degenerative valve disease (Fuentes *et al.*,2002, Boyle and Leech, 2012, Summerfield *et al.*,2012, Haggstrom *et al.*, 2013, Boswood *et al.*2016, Vollmar and Fox, 2016). However, the use of pimobendan remains controversial in cats with hypertrophic cardiomyopathy as left ventricular outflow tract obstruction (LVOTO) is a concurrent complication in 47.5 -67% cats with hypertrophic cardiomyopathy (Pyne *et al.*,2010). It has been shown that pimobendan imparts a significant benefit in left auricle (LA) function in cats with hypertrophic cardiomyopathy (HCM) and does not significantly increase systolic function and LOVT obstruction (Oldach *et al.*, 2019). Pimobendan (0.25 mg/kg IV) has shown positive chronotropic and inotropic effects in healthy horses (Afonso *et al.*,2016), but information regarding its use in the management of heart failure in horses is lacking.

H. Vasodilators

Hydralazine

It relaxes smooth muscles in systemic arterioles but do not have effect on systemic venous tone. It is a potent arteriolar dilator. It also causes hypotension. Hydralazine should be reserved for the cases that remain refractory to ACE-inhibitors or dogs having acute heart failure owing to mitral regurgitation. It can be used at a starting dose of 1.0 mg/kg PO BID in dogs not receiving ACE-inhibitors. The dose can be titrated upward up to 1.8 mg/kg – 2.8 mg/kg if there is poor or no response. Blood pressure monitoring should be done in dogs undergoing hydralazine therapy. Use of hydralazine in dog already receiving ACE- inhibitors should be done cautiously.

Hypotension, anorexia, vomiting and diarrhea are the side effects of hydralazine. On development of hypotension and other symptoms, the dose of hydralazine should be reduced or discontinued. Reflex sympathetic tachycardia can also occur in dogs with heart failure. If reflex tachycardia develops, it may be treated with beta blockers or digoxin.

Nitrates

Nitroglycerine and nitroprusside fall in this category. Nitrates relax vascular smooth muscle and are the drugs for systemic venodiltion in canine and feline with heart failure. Nitrates are well absorbed from the gut and are rapidly metabolized by the liver. Orally administered nitrates have low bioavailability. Nitroglycerin is used transdermally in humans, dogs and cats with heart failure. It can be used topically on the ears. It is generally used in conjunction with furosemide. It is available as 2% Ointment/cream and can be applied on the skin @ 1.0 inch /kg body weight twice daily in dogs or 1/8th to 1/4th inch TID/QID in cats. The cream or ointment should be applied on hairless areas using gloves.

Nitroprusside

Nitroprusside is sodium nitroferricyanide. It releases nitric oxide in vascular smooth muscle and is a potent venidilator and arteriolar dilator. It is recommended intravenously for short term treatment in severe or fulminant heart failure. Its half-life is few minutes as it is rapidly metabolized after intravenous injection. It is indicated in the management of acute mitral regurgitation also. Its doses are variable and is used from 2 to 25 microgram/kg. Sodium nitroprusside should be protected from light, heat and moisture.

Hypotension and cyanide toxicity are the adverse effects of sodium nitroprusside.

Isosorbide Dinitrate

It is an organic nitrate being used orally. It can be used @0.5 to 2mg/kg orally twice daily. It is commonly used sublingually in the management of acute angina and heart failure and also orally for the long term management of angina and heart failure in humans. It is also used intravenously in humans for the management of unstable angina and heart failure. It is also recommended intracoronary for percutaneous transluminal coronary angioplasty.

I. Sympathomimetics

Sympathomimetics increase contractility of heart muscles, conduction speed as well as heart rate. These drugs can stimulate both alpha and beat adrenergic receptors. Dopamine and Dobutamine are sympathomimetic of choice. These are less arrhythmogenic and produce small increase in heart rate and are suitable for use in the management of heart failure. They have short half-life and generally used as a constant rate infusion. Details are given under the heading inotropes.

J. Bipyridine Compounds

These compounds are also having positive inotropic effect and can increase myocardial contractility with mild systemic arteriolar dilatation. This group has two drugs viz. amrinone and milrinone. Of these milrinone is more potent (around 30 times). Both drugs can be used orally.

Amrinone

Amrinone is a bipyridine compound that inhibits phosphodiesterase III. It causes vasodilation. In higher doses it not only causes vasodilation but also reduces blood pressure and increases heart rate. For sustained effects, amrinone is to be given as constant rate infusion (CRI) . Though it has wide margin of safety, it can exacerbate ventricular arrhythmias. It is used for short inotropic support.

Milrinone

Milrinone, related to amrinone, is also a selective inhibitor of phosphodiesterase III. It is more potent than amrinone. It is effective inotrope agent both in normal dogs and dogs with heart failure. It has been used in the management of heart failure in dogs. It exacerbates ventricular tachyarrhythmia in some cases. It can be given orally @0.5 to 1.0 mg/kg.

K. Anxiolytic Drugs

These drugs are used to reduce anxiety. Severe respiratory distress in cardiac compromised dogs and cats may lead to anxiety. In humans with heart failure anxiolytic drugs are given to reduce stress in the patients..

Morphine

Morphine is commonly used anxiolytic in humans with respiratory distress secondary to heart failure .It has venodilating property also. It is not recommended in cats. In dogs it can be used @ 0.1-0.2 mg/kg SC. Dose can be repeated as per need to get desired result. Adverse effect of morphine is respiratory depression.

Acepromazine

It is a tranquilizer of phenothiazine group. It can be used to sedate dogs with respiratory distress owing to heart failure. It does not depress respiration but reduces agitation. Because of its alpha adrenergic blocking property, it decreases peripheral vascular resistance and can be used in dogs with mitral regurgitation. It can be used @ 0.01 to 0.2 mg/kg IV /IM. It should be avoided in boxers because of severe hypotension and bradycardia. It should not be used in conjunction with opioids.

L. Bronchodilator Therapy

Though theophylline and aminophylline have occasionally been used in humans with heart failure, their utility is not without question. There is no specific indication for their use in dogs with heart failure. Both drugs are nonspecific phosphodiesterase inhibitor, and have very weak positive inotropic and diuretic effects. But their use in dogs with heart failure has no added advantage.

M. Cough Syrups

Cough syrups in dogs with coughing secondary to pulmonary edema due to heart failure, are contraindicated. Coughing due to compression of airways owing to enlarged left atrium may respond to hydrocodone bitartrate with homatropine and butorphanol. Dextromethorphan may also have some beneficial effect.

Cardiac Drugs with Doses - At a Glance

1. **Anti-arrhythmic Drugs-**These drugs are indicated to restore cardiac rhythm and rate.

a. Bradyarrhythmias

Atropine sulfate @0.04 mg/kg SC /IV for both dogs and cats

Dopamine @2 to 10 µg/kg/min IV (not more than 3 days)

Glycopyrrolate @0.05 to 0.01 mg/kg SC / IM

Propantheline @0.5-1.5 mg/kg PO TID for dogs/

Terbutaline@1.25 to 2.5 mg PO BID/TID

Pacemaker implantation

b. Supraventricular tachycardia

Atenolol @ 6.25-25.0 mg PO OD/BID

Digoxin: See positive inotropes

Diltiazem @ 0.5-1.5 mg/kg PO TID

@0.1-0.25 mg/kg IV over 2 minutes

@ 7.5 mg/cat PO TID

Esmolol@0.1-0.5 mg/kg IV rapidly

Propranolol @5.0- 40.0 mg PO TID

@ 0.1 - 0.3 mg/kg IV slowly

Verapamil @1.0- 5.0 mg/kg PO TID

@ 0.05 to 0.25 mg/kg IV slowly

c. Ventricular Tachycardia

Amiodarone@ 10.0-20.0 mg/kg OD for 7-10 days

Followed by 5 mg/kg at 48h interval or 15mg/kg OD.

Atenolol- as in supraventricular tachycardia

Esmolol- as in supraventricular tachycardia

Lidocaine@2.0- 4.0 mg/kg IV.It can be repeated up to 8 mg/kg over 20 minutes

@ 0.2- 0.75 mg/kg IV as bolus . It can be repeated 1-2 .

Procainamide@ 5.0 to 15.0 mg/kg PO TID

@ 5.0- 10.0 mg/kg IV

Propranolol-as in supraventricular tachycardia

Quinidine@5.0- 15.0 mg/kg PO QID

Sotalol @ 1.0-2.0 mg/kg PO BID.

Tocainide@5.0-10.0 mg/kg PO TID

2. **Arterial Vasodilators**-These drugs are indicated to decrease afterload.

   Amlodipine @0.1-0.2 mg/kg PO OD/BID

   Benazepril @ 0.25-0.50 mg/kg PO OD

   Captopril@ 0.5 to 2.0 mg/kg PO TID

   Diltiazem@ 0.1-0.2 mg/kg IV slowly

   @ 0.5-1.5 mg/kg PO TID

   @ 7.5 mg/cat PO TID

   Enalapril@0.25 -0.50 mg/kg PO OD/BID

   Hydralazine @1.0- 3.0 mg/kg. PO BID

   Prazosin@1.0 mg PO TID (up to 15 kg dogs)

   @2 .0mg PO TID (>15 kg dogs)

   Sildenafil@ 0.5-1.0 mg/kg PO OD/BID

   Sodium nitroprusside@1.0- 5.0μg/kg/min IV

3. **Diuretics**- These drugs are indicated to reduce excessive preload and fluid retention.

   Chlorthiazide@20.0- 40.0 mg.kg PO BID

   Furosemide @1.0- 4.0 mg/kg IV/IM/SC /PO OD/TID

   @ 0.66 mg/kg/min CRI

   Cats @1.0-4.0 mg/kg IV/SC/PO OD/TID

   Hydrochlorothiazide @2.0- 4.0 mg/kg PO OD/BID

   Spironolactone @ 0.50-2.50 mg/kg PO BID

   Torsemide@ 0.2 mg/kg PO OD/TID or one tenth of furosemide dose.

4. **Positive Inotropes**-These drugs are used to increase myocardial contractility.

   Amrinone@1.0- 3.0 mg/kg IV followed by 10 to 100 μg/kg/min CRI

   Digoxin @0.01-0.02 mg/kg PO initial on 1st day.

   @ 0.005 - 0.01 mg/kg PO BID for maintenance.

   @ 001 to 0.02 mg/kg rapid IV. Half of total IV dose immediately and one fourth IV at 30-60 min.interval as per need.

Dobutamine@ 1.5 – 20.0μg/kg/min IV for less than 72 hours

Dopamine @ 2.0- 10.0 μg/kg/min IV for less than 72 hours

Pimobendan@ 0.25 mg/kg PO BID empty stomach

5. **RAAS Inhibitors**-These drugs are used to reduce neurohormonal aberrations.

   Benazepril@ 0.25-0.50 mg/kg PO OD

   Cats @ 0.25 – o.5 mg/kg PO OD

   Captopril @0.5 – 2.0 mg/kg PO TID

   Enalapril @0.25-0.5 mg/kg PO BID

   Spironolactone @ 0.5 mg/kg/ PO OD

5. **SNS Blunting**-These drugs are used to reduce neurohormonal aberrations.

   Atenolol @ 0.25-1.0 mg/kg PO

   Carvedilol@ 0.1-0.2 mg/kg PO OD increasing to 0.5-1.0mg/kgPO BID over 6 wks

   Digoxin@ 0.005 - 0.01 mg/kg PO BID maintenance dose.

   Propranolol @ 5.0- 40.0 mg PO TID

6. **Venodilators**-These drugs are used to reduce excessive preload and fluid retention.

   Benazepril @ 0.25-0.5 mg/kg PO OD

   Captopril@ 0.5 – 2.0 mg/kg PO TID

   Enalapril@0.25 mg/kg PO BID

   Nitroglycerin 2% ointment @1/4 inch/5kg topical TID for 1st One day.

   Prazosin@1.0 mg PO TID (upto15 kg weight)

   @ 2.0 mg PO TID (>15 kg weight)

   Sodium nitroprusside@1.0 – 5.0μg/kg/min IV CRI

   Triamterene@ 2.0 to 4.0 mg/kg/ PO OD

7. **Bronchodilators**

   Terbutaline @ 2.5- 5.0 mg /dog PO divided in three doses per day.

Or 0.01 mg/kg SC 4hrly.

@ 1.5 mg/cat P O divided three doses per day. Or 0.01 mg/kg

SC 4 hrly.

Theophylline ER @ 10 mg/kg PO BID for dogs and cats

8. **Sedatives**

Acepromazine @ 0.005-0.01 mg/kg IV/SC once .

Butorphanol @ 0.05-0.3 mg/kg IV/SC for dogs and cats. Can be repeated as per requirement.

Buprenorphine @ 0.005- 0.02 mg/kg IV/SC TID/QID for both dogs and cats.

Fantanyl @2.0-3.0 microgram/kg IV followed by 1.0-5.0 microgram/kg/h CRI for both dogs and cats

9. **Anti-thrombitics**

Aspirin @ 5.0 mg/cat OD with food.

Dalteperin @ 100.0-150 IU/kg SC BID for cats

Enoxaparin@ 1.0 mg/kg SC BID for cats

UF Heparin @ 100U/kg IV once then 200 U/kg SC TID for cats

Tissue Plasminogen Activator @ 00.75 mg IV as bolus ,then 2 5 mg IV over 30 minutes followed by 1.75 mg IV over 1 hr (total dose 5 mg) for cats.

Remarks

1. Doses given above are taken from the published papers from the literature. There may be some variations in doses as different authors have recommended a little different doses.
2. The drug doses are mainly for dogs if there is no specific mention for cat.

Cardiac Drugs and their Commercial Preparations

Generic Name of Drugs	Commercial Preparations with their presentations. Manufactures names are given in bracket.
Acepromazine	Acepro 25 (Bimeda-MTC, Animal Health, INC). Supply by GNH India Inj 25 mg/ml Acepromazine maleate (Phoenix) Inj 10 mg/ml
Amiodarone	Aldarone (Alidac) Inj. 50mg/ml ,3 ml ampoule Amiodarone (Shree Ganesh) Inj. 50 mg/ml , 3 ml ampoule Duron (Samarth Inj. 50 mg/ml , 3 ml ampoule Amiodar 100 (Micro Lab) Tab 100 mg, 200 mg Eurythmic (Troikaa) Tab 100 mg, 200 mg, Inj. 50 mg/ml, 3ml Ampoule
Amlodipine	AM-5 (Orchid) Tab 5 mg Amdepin (Cadila Pharma) Tab 5, 10 mg Amlocor (Torrent) Tab 2.5, 5, 10 mg Amlogard (Pfizer) Tab 5, 10 mg Amlong (Micro Labs) Tab 2.5, 5, 10 mg Amlopin (USV) Tab 2.5,5, 10 mg Amlopres (CIpla) Tab 2.5, 5, 10 mg Amtas (Intas) Tab 2.5,5,10 mg Corvadil (Unichem) Tab 2.5,5,10 mg Stamlo (Dr. Reddy's) Tab. 2.5,5,10 mg. Emadine (E-merck) Tab 5, 10 mg
Amrinone	Amicor (Samarth Pharma) Inj. 5 mg/ml 20 ml injection Cardiotone VHB) Inj 5 mg/ml 20 ml injection

Asprin (Acetyl Salicylic Acid)	ASA (German Remedies) Tab 50 mg Aspent (Solus) Tab 60 mg Cotaspirin (Natco) Tab 50,500 mg Delisprin (Aristo) Tab 75, 150 mg E-Prin (Elder) Tab 150 mg Loprin -75 (Unisearch) Tab 75 mg Loprin-DS (Unisearch) Tab 162.5 mg Optaz-75 Tab 75 mg
Atenolol	Atcardil (Sun Pharma) Tab 50,100 mg Atcom (Comed) Tab 25,50 mg Atelol (Themis Pharma0 Tab 25,50,100 mg Atenova (Lupins) Tab 25,50,100 mg Beta (Stadmed) Tab 25,50,100 mg Betacard (Torrent) Tab 25,50,100 mg BP-nol (Elder) Tab 25,50,100 mg Lonol (Khandelwal) Tab 25,50,100 mg Tenolol (Ipca) Tab 25,50,100 mg Inj.0.5mg/ml 10 ml injection
Atropine	ATP (Biomilcron) Amp. 0.6 mg/ml Atro (Prem Pharma) 100 ml vial Tropine (Neon Labs) 0.6 mg/ml 10 ml vial Atropine Sulphate (Mnorvel) 1.0mg/ml 100 ml vial1.1
Benazepril	Aceptor (Sava Vet) Tab 2.5,5.0, 10.0 mg Benace (Novartis) Tab 5, 10 mg

Buprenorphine	Addnok (Rusan H. Care) Tab 0.2 mg. Buprine (Paksons) Inj.0.3mg/ml, 2ml ampoule Buprinoe (Astra-Zeneca) Inj.0.3 mg/ml, 2ml ampoule Norphin (Unichem) Inj.0.3 mg/ml, 2 ml ampoule Pantorel (Khandelwal) Inj. 0.3 mg/ml, 1 ml/ 2ml ampoule Tidigesic (Sun) Inj.0.3 mg/ml, 1 ml/ 2 ml ampoules Tab 0.2 mg
Butorphanol Tartrate	Butodol (Neon Lab) Inj. 1mg/ml, 2 mg/ml, 5 ml injection Butrum (Aristo) Inj 1mg/ml, 2mg/ml, 5 ml injection
Captoril	Acetin (Wockhardt) Tab 25 mg Angiopril (Torrent) Tab 25 mg Angiopril-DU (Torrent) Captoril 25 mg+ Hydrochlorothiazide 15 mg Capotril (Lupin) Tab 12.5 , 25 mg
Chlorthiazide (Hydrochlorothiazide)	Aquazide(Sun) Tab. 12.5 mg Bpzide (Stadmed) Tab. 12.5, 25 mg Hydrazide (Cipla) Tab 12.5 mg Hydride (Micro Lab) Tab 12.5 mg
Carvedilol	Carca (Intas) Tab 3.125, 6.250, 12.5 mg Cardivas (Sun Pharma) Tab 3.125, 6.25, 12.5, 25 mg Carloc (Cipla) Tab 3.125, 6.25, 12.5, 25 mg Carvil (Zydus Medica) Tab 3.125, 6.25, 12.5 mg Caslot (Stanecare) Tab 3.125, 6.25, 12.5, 25 mg CVL (Themis Medicare Tab 3.125, 6.25, 12.5

Coenzyme Q10	UB-10 (Bioplasm) Cap 100 mg Miraqule (Alkem) Soft gel 50, 100 mg
Dalteparin (Low molecular weight heparin)	Fragmin (Pharmacia & Upjohn) PF syringe 2500 IU, 5000 IU Inj 10000IU, 100000 IU
Digoxin	Cardioxin (Novartis) Tab 0.25 mg Inj 0.5 mg/ 2ml Digox (Cadila Pharma) Tab 0.25 mg Dixin (Samarth) Tab 0.25 mg Inj. 0.5 mg/2ml Syrup 1.5mg/30 ml Lanoxin (Glaxo-Smithkline) Tab 0.25 mg
Diltiazem	Angizem (Sun Pharma) Tab 30, 60 mg Cardem (Anglo French) Tab 30,60 mg Channel (Micro Labs) Tab 30,60 mg Channel SR (Micro Labs) Tab 90,120 mg Dicard (Intas) Tab 30,60, 90 mg Dilcal (Nicholas Piramal) Tab 30,60 mg Tab 90 mg SR Dilgard (Protec) Tab 30,60 mg Dilgard XL (Protec) Tab 90,120,180 mg Dilgina (Kopran) Tab 30 mg Dil Time (Zydus Medica) Tab 30,60 mg SR Tab 90,120 mg Dilzem (Torrent) Tab 30,60 mg Inj 5 mg SR tab 90 mg Masdil (Lupin) Tab 30,60 mg Masdil-BD (Lupin) Tab 90,120, 180, 240 mg

Dobutamine	Cardiject (Sun Pharma) Inj 50 mg/4ml 250 mg/20 ml Dobunes-Lypho (Leechem Biotech) Inj 250 mg/vial Dobutasys (Sympar) Inj 250 mg, 5 ml vial Dobutrex (Nicholas Piramal) Inj 250 mg/vial.
Dopamine	Dopacard (Nicholas Piramal) Inj 200mg/5ml Dopamine (TTK) Inj 200 mg/5ml Dopinga (Inga Laboratories) Inj 40 mg/ml, 5 ml ampoule
Enalapril	Bol (Cadila Healthcare) Tab 2.5,5. 10 mg Converten (Khandelwal) Tab 2.5,5,10 mg Enace (Nicholas-) Tab 2.5,5,10 mg Enam (Dr. Reddy's) Tab 2.5,5,10 mg Enapril (INtas) Tab 2.5,5,10,20 mg Enpril (Wockhardt) Tab 2.5,5,10 mg Envas (Cadila Pharma) Tab 2.5,5.10,20 mg Normace (Unichem) Tab 2.5,5,10 mg Oren (Natco) Tab 2.5,5,10,20 mg Vasonorm (Kopran) Tab 2.5,5,10,20 mg
Enoxaparin	Clexane (Sanofi Aventis) Inj 20 mg/0.2 ml 40 mg/0.4 ml H-Parin (Uniword Marketing) Inj 20,40 60 mg Leeparin (Leechem Biotech) Inj 40, 60 mg Lomonox (Neiss Labs) Inj 40,60 mg LMWX (Nicholas) Inj 20,40,60, 80 mg

Epinephrine	Adrenaline (Keshav Das and Sons) Inj Epinephrine 1 mg/ml, ampoule Adrenaline HCL (Sarvasya Trading (India) Private ltd) Inj Epinephrine 1 ml ampoule Adrenalin (Maya Biotech Pvt Ltd) 1mg 1:1000 1ml vial Epinealth (Livealth Biopharma Pvt Ltd) 1 mg/ml ampoule
Esmolol	Miniblock (USV) Inj 10 mg/ml 10 ml vial Inj. 25 mg/ml 10 ml vial
Fentanyl	Durogesic (Johnson and Johnson) TD-patch 25,50,75 mcg/h
Furosemide	Remot-Vet (Morvel Lab) Inj 50 mg/ml Ridema (Vetoquinol) Inj 50 mg/ml Lasix (Sanofi Aventis) Inj 10 mg/2 ml ampoule Tab 40 mg Exna-K Micro Lab) Tab 40 mg Frusis (Gnosis Pharma) Tab Furosemide 20 mg+ Spironolactone 50 mg
Glycopyrrolate	Glyco-P (Khandelwal) Inj 0.2 mg/ml 1ml ampoule Pyrolate (Neon lab) Inj. 0.2 mg/ml, 1 ml ampoule and 10 ml Vial
Heparin	Beparine (Biological E) Inj 1000, 5000 IU/ml Caprin (Samarth) Inj 5000, 25000 IU
Hydralazine	Nepresol (Novartis) Tab 25 mg
Isoproterenol	Isuprel (Cipla) Inj 0.2 mg/ml
Isosorbide dinitrate	Anzidin (Stadmed) Tab 5, 10 mg Cardicap-TR (Sun Pharma) Cap 20,40 mg Dinospray (Cipla) Spray 200 metered doses Isordil (IPCA) Tab 5, 10 mg

	Sorbitrate (Nicholas) Tab 5, 10 mg
Lidocaine/ Lignocaine/Xylocaine	Gesicain (SPPL) Inj. 2% 30 ml vial Lox-2% (Neon Lab) Inj. 21.3 mg/ml, 30 ml vial Xylocard (AstraZeneca) Inj.21.3 mg/ml, 50 ml vial
Lisinopril	Acinopril (Nicholas Piramal) Tab 2.5,5.0, 10.0 mg Cipril Cipla) Tab 2.5, 5.0, 10.0 Cipril-H (CIpla) Lisinopril 2.5 m+ Hydrochlorthiazide 12.5 mg Inras (Merind) Tab 2.5,5.0 10.0 mg Lipril (Lupin) Tab 2.5,5.0, 10.0 mg Lisoril (IPCA) Tab 2.5,5.0.,10.0 mg Lisotec (Sun) Tab 2.5, 5.0 mg Nivant (German Remedies) Tab 2.5,5.0.10.0, 20.0 mg Sinopril (Themis) Tab 5 mg
Mexiletine HCL	Mexitil (German remedies) Cap 50,150 mg Inj. 250 mg
Milrinone	Primacor (Sanofi Aventis) Inj 10 ml (1 mg/ml)
Morphine	Duramor (Bio E) Tab 10,30 mg Morphine Sulphate (Haffkine) Inj 10 /ml
Nifedipine	Adalat-Ors (Bayer) Cap 30 mg (extended release) Tab 10,20 mg (Retard) Angiblock (Alkem) Cap 10 mg Tab 10,20 mg (retard) Calcigard (Torrent) Cap 5,10 mg Tab 10,20 mg (retard) Cardules (Nicholas) Cap 10 mg Tab 10,20 mg (retard)

	Depin (Zydus Cadila) Cap 5, 10 mg Tab 20 mg (retard) Nifedine (Nicholas Piramal) Cap 5, 10 mg SR 10,20 mg Nifelat (Cipla) Cap 5, 10 mg Tab 20 mg (retard) Niferil (U.S. Vitamins) Tab 20 mg (retard)
Nitroglycerine	Lygly (Cu-card) Tab 2.6, 6.4 mg Nitrocin (Samarth) Inj 25, 50 mg Nitrocure (Panacea) Inj 25 mg/5 ml Myovin (Cadila) Ointment 2% Nitroderm TTS (Novartis) TTS 5 (25 mg)
Nitroprusside (Sodium Nitroprusside)	Nipress (Samarth Pharma) Inj 50 mg vial Niside (Shree Ganesh Inj 50 mg vial Sonide (Gufic) Inj 50 mg vial
Phenytoin	Epsolin (Cadila Health Care) Inj.50 mg/ml 2 ml ampoule
Pimobendan	Safe Heart (Sava Vet) Tab.1.25,5.0, 10.0 mg. Cardiopet (Corise Canine care) Tab. 2.5, 5.0 mg Pimoben solo (Vivaldis Animal Health) Tab.1.25, 5.0 mg Pimocard (Swiss Biocare) Tab. 2.5 mg, 5.0 mg Pimoben (Vivaldis) Tab Pimobendan 1.25mg + Spironolactone 5 mg Tab Pimobendan 2.5 mg + Spironolactone 10 mg
Propantheline Bromide	Probanthine (RPG) Tab 15 mg

Propranolol	Betabloc (USV) Tab 10, 40 mg Ciplar Cipla) Tab. 10, 40 mg Inderal ICI) Tab 10,40, 80 mg Propal (Sigma) Tab 10, 40 mg Propranolol (Samarth Pharma) Inj. Vial 1mg/ml
Prazosin	Minipress XL GITS (Pfizer) Tab. 2.5, 5.0 mg Prazopress (Sun Pharma) Tab 1, 2 mg Prazovin-XL 5(Bison Biotec) Tab 5 mg
Quinidine	Natcardine (Franco-Indian) Tab 100mg Quinidine (Burroughs Wellcome) Tab 200 mg
Ramipril	Alraxpril (Alpha Pharma) Tab 2.5, 5 mg Athpril (Athens Lab) Tab 5 mg Bampril (Biotec) Tab 2.5, 5 mg Cardace (Sanofi Aventis) Tab 1.25, 2.5, 5, 10 mg Cardiopril (Dr Reddy's Lab) Cap 1.25, 2.5, 5, 10 mg Kapril (KAPL) Tab 2.5,5 mg Proace (Alembic) Tab 2.5,5, 10 mg Ramace (Astra Zeneca) Cap 1.25,2.5,5,10 mg Ramipres (Cipla) Cap 1.25,2.5,5,10 mg Topril (Torrent Pharma) Cap 1.25, 2.5,5 mg Zorem (INtas) Cap 1.25,2.5, 5 mg
Sildenafil	Alsigra (Alembic) Tab 25, 50 mg Androz (Torrent) Tab 25, 50,100 mg Caverta (ranbaxy) Tab 25,50,100 mg

	Edegra (Sun Pharma) Tab 25,50,100 mg Enthusia (Lupin) Tab 50,100 mg Juan (Cadila) Tab 25,50,100 mg Manforce (mankind) Tab 50,100 mg Rezum (Dr. reddy's) Tab 50 mg Silagra/Suhagra (Cipla) FC Tab 25,50,100 mg Viraha (Micro Labs) Tab 25,50 mg
Sotalol	Sotagard (GlaxoSmithkline) Tab. 40, 60 mg
Spironolactone	Aldactone (RPG) Tab 25,100 mg Fruselac (Lupin) Spironolactone 50 mg+ Furosemide 20 mg Lasilactone 50 (Sanofi Aventis) Spironolactone 50 + Furosemide 20 mg Spiromide (RPG) Spironolactone 50 + Furosemide 20 mg
Terbutaline	Asmaril (DWD) Syrup Bricanyl (Astra Zeneca) Tab 2.5, 5 mg Syr 1.5 mg/ 5 ml Inj 0.5 mg/ml Terbet Forte (Strassenburg) Tab 5 mg Tetrasma (Medopharm Tab 5 mg
Theophylline	Broncordil (USV) Tab SR 200 mg Elixir 80 mg/ml Theo PA (GSK) Tab 100,300 mg Theo SR (Alembic) Tab 200 mg Theobid (CIpla –Protec) Tab SR 200, 300 mg Theoday (Cipla-Protec) Tab SR 400,600 mg

	Theoped (Cipla –Protec0 Syr 50 mg/5ml Theophylline (Intra Lab) Tab 400 mg TSR (Bestochem) Tab 200,400 mg
Tissue Plasminogen Activator	Actilyse (German remedies) Inj 50 mg with 50 ml solvent water (50 ml vial)
Triamterene	Ditide (GSK) Tab benzthiazide 25 mg + Triamterene 50 mg
Tocainide	Tonocard (AstraZeneca), Tocolol, Tocain, Mexitil Tab 400, 600 mg
Torasemide	Cu-Tros-10 (CU-Card) Tab 10 mg Diurator (Ellife-Elder) Tab 10.20 mg Dytor (Cipla) Tab 10.20,100 mg Inj. 2 ml ampoule Metor (Globus Remedies) Tab 5, 10 mg Torsemi (FDC) Tab 10 mg.
Verapamil	Calaptin (Boehringer Mannheim) Tab 40, 80 mg S-R Tab 120, 240 mg Inj. 5 mg/2 ml VPL (Samarth) Inj. 5mg/2 ml Vasopten (Torrent) Tab 40,80,120 mg Veramil (Themis Chemicals) Tab 40, 80 mg

References

Afonso,T., Giguere,S.,Rapoport,G.,Barton,M.H. and Coleman, A.E. (2016).Cardiovascular effect of pimobendan in healthy mature horses. Equine Vet.J.48:352-356.

Boswood, A., Haggstrom, J., Gordon, S.G., Wess, G., Stepien, R.L., Oyama, M.A., Keene, B.W., Bonagura, J., Mac Donald, K.A., Patteson, M., Smith, S., Fox, P.R., Sanderson, K., Wooley, R., Szatmari, V., Menaut, P., Church, W.M., O'Sullivan, M.L., Jaudon, J.P., Kresken, J.G., Rush, J., Barrett, K.A., Rosenthel, S.L., Saunders, A.B., Ljungvall, I., Deinert, M., Bomassi, E., Estrada, A.H., Fernandez Del Palacio,M.J., Moise,N.S., Abbott, J.A., Fujii, Y., Spier, A., Luethy, M.W., Santilli, R.A., Uechi, M., Tidholm, A.

and Watson, P. (2016). Effect of pimobendan in dogs with preclinical myxomatous mitral valve disease and cardiomegaly: the EPIC study-a randomized clinical trial. J. Vet ..Int. Med.30:1765–1779.

Boyle, K.L. and Leech, E. (2012). A review of the pharmacology and clinical uses of pimobendan. J. Vet.Emerg. Critic. Care 22:398–408.

Chu, K.M., Shieh, S.M. and Hu, O.Y. (1995). Pharmacokinetics and pharmacodynamics of enantiomers of pimobendan in patients with dilated cardiomyopathy and congestive heart failure after single and repeated oral dosing. Clin .Pharmacol. Ther.57:610-621.

Fuentes, V.L.(2004). Use of pimobendan in the management of heart failure. Vet. Clin. North Am. Small Anim. Pract.34:1145-1155.

Fuentes, V.L., Corcoran, B., French, A., Schober,K.E., Kleemann, R. and Justus C. (2002).A double-blind,randomized, placebo-controlled study of pimobendan in dogs with dilated cardiomyopathy. J Vet Int Med. 16:255–261.

Giselle,H. (1990). Pharmacologic features and physiological effects of dopamine. J.Am. Vet. Med. Assoc. 9: 1209-1211.

Goldberg, L.I. (1972). Cardiovascular and renal actions of dopamine: Potential clinical application.Pharmacol. Rev. 24:1-29.

Goldberg,L.I. and Raifer,S.I. (1985). Dopamine receptors: applications in clinical cardiology. Circulation 72:245-248.

Hagemeijer, F.(1993). Calcium sensitization with pimobendan: pharmacology, haemodynamic improvement, and sudden death in patients with chronic congestive heart failure. *Eur. Heart J.* 14:551-566.

Haggstrom, J., Boswood, A.,O'Grady, M., Jons, O., Smith, S., Swift, S., Borgareilli, M., Gavaghan, B., Kresken, J.G., Patteson, M., Ablad, B., Bussadori, C.M., Laus, G. Kovacevic, A., Rapp, M., Santilli, R.A., Tidhlm, A., Riksson, A., Belanger, M.C., deinert, M., Little, C.J.L., Kvart, C., French, A., Ronn-Landbo, M., Wess, G., Eggertsdottir, A.,O'Sullivan, L.M., Schneider, M., Lombard, C.W., Dukes-McEvan, J., Willis, R., Louvet,A. and Difruscia,R.(2013). Longitudinal analysis of quality of life, clinical, radiographic, echocardiographic, and laboratory variables in dogs with myxomatous mitral valve disease receiving pimobendan or benazepril: the QUEST study. J .Vet .Int. Med. 27:1441–1451.

Kumar, A. and Varshney, J.P. (2005). Atrial standstill in a German shepherd dog. Indian Vet. med. J. 29:225-226.

Lefbon, B.K.(2005). Sildenafil and novel cardiovascular therapies, in *Proceedings*. Am. Coll. Vet. Intern. Med. 2005 :75-80.

Lewis, B.S. and Bruce, L.G (1988). The renal response to low dose dopamine. J. Surg. Res. 45:574-588.

Oldach, M.S., Ueda, Y., Ontiveros, E.S., Fousse, S.L. Harris, S.P. and Stern, J.A. (2019). Cardiac effects of a single dose of pimobendan in cats with hypertrophic cardiomyopathy: A randomized placebo controlled crossover study. Front. Vet.Sci., 04 February 2019. Heeps://doi.org/10.3389/fvets.2019.00015.

Opie, L.H., Frisman, W.H. and Thadani U. (1995).Calcium channel antagonists (calcium entry blockers). In: Drugs for the Heart. Opie,L.H. (ed), 4th ed. Philadelphia, PA: WB Saunders.pp:50-82.

Payne, J., Luis Fuentes, V., Boswood, A., Connolly, D., Koffas,H. and Brodbelt, D.(2010). Population characteristics and survival in 127 referred cats with hypertrophic cardiomyopathy (1997 to 2005). J .Small Anim.Pract. 51:540–547.

Summerfield, N.J., Boswood, A., O'Grady ,M.R., Gordon, S.G., Dukes-McEwan, J., Oyama, M.A., Smith, S., Patteson, M., French, A.T., Culshaw, G.J., Braz-Ruivo, L., Estrada, A., O'Sullivan, M.L., Loureiro, J., Wills, R. and Watson, O. (2012). Efficacy of pimobendan in the prevention of congestive heart failure or sudden death in Doberman Pinschers with preclinical dilated cardiomyopathy (the PROTECT Study).J. Vet.Int.Med. 26:1337–1349.

Vollmar, A.C. and Fox, P.R.(2016). Long-term outcome of Irish wolfhound dogs with preclinical cardiomyopathy, atrial fibrillation, or both treated with pimobendan, benazepril hydrochloride, or methyldigoxinmonotherapy. J. Vet. Int. Med. 30:553–559

Section 3
Heart Failure in Other Animals and Man

10

Heart Failure in Feline

Heart diseases in feline are silent killer as cats are able to hide their symptoms in early stage. Cats are brought to clinics only when disease is more advanced. Heart diseases are different from heart failure. Heart failure is the end stage of heart diseases.

- All heart diseases may not lead to heart failure but heart failure is definitely due to heart disease.
- The major heart disease in cats are hypertrophic cardiomyopathy, valvular defects, and birth defects.
- Development of clinical signs is generally slow and progressive, but cats may also be presented with acute illness.
- Congestive heart failure or aortic thromboembolisms are the two main clinical manifestations.
- Heart failure is a syndrome owing to severe cardiac dysfunctions and broadly classified as left heart congestive failure or right heart congestive failure.
- Hypertrophic cardiomyopathy is the most common cause of heart failure in cats (Goutal *et al.*, 2010). Heart failure is also associated with dilated cardiomyopathy, restrictive cardiomyopathy, arrhthymogenic right ventricular cardiomyopathy, congenital heart diseases, hyperthyroidism and anemia.
- Smith *et al.* (2004) have reported a few cases of congestive heart failure in cats after administration of various glucocorticoids. Several cats succumbed to it and others had reversible left ventricular hypertrophy. It was suggested that glucocorticoid administration resulted in a hyperosmolar increase in plasma volume which could account for congestive heart failure. But changes in cardiac chamber dimension were not documented.

- Recently transient myocardial thickening (TMT) associated with catecholamine toxicity has been reported to cause heart failure in cats (Novo Matos, *et al.,* 2017). TMT is echocardiographically similar to hypertrophic cardiomyopathy.
- A report has recently described a case of refractory congestive heart failure in a five year old male obese domestic short hair cat associated with SARS-CoV-2 associated myocarditis in France (Chetboul *et al.,* 2021). The authors reported that the cat was sero positive for SARS-CoV-2 and had high cTn-I level.
- Heart failure in cats is mainly diastolic heart failure
- Congestive heart failure in cats is manifested by tachypnea, open mouth breathing, bluish gums, pulmonary edema, ascites, pleural effusions (Fig . 31) and pericardial effusions; cool extremities (paws, ears); decreased appetite or anorexia; and reluctance to exercise. Accumulation of fluid in abdomen, pleura, pericardium, and limb edema and in liver is the reflection of right heart failure. Right heart failure in cats is less common. While left heart failure is more common in cats. Aortic thromboembolism is manifested by functional defect of the affected hind limb. Initially thromboembolism is painful. The affected limbs are cold to touch and there is no arterial pulsation. Limbs becomes weak with loss of sensation.

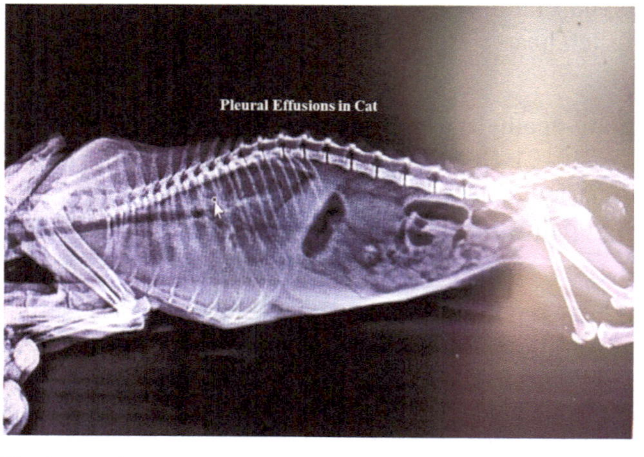

Fig. 31: Right lateral radiograph of a cat showing pleural effusions and enlarged liver.

- Oxygen and furosemide are the corner stone of the therapy for stabilizing cats with acute heart failure.

- Oxygen therapy can be provided through oxygen masks, oxygen cages or flow by. Efforts should be made to minimize stress to cats while providing oxygen therapy.
- In acute cases furosemide gives better result if given intravenously (1.0-2.0 mg/kg as bolus or CRI). Sometimes attempting intravenous injection may be disastrous in cats with severe dyspnea. Therefore, in such cases furosemide may be given intramuscularly initially. Because of poor peripheral perfusion in cats with heart failure, it is better to avoid subcutaneous administration of furosemide in cats. Furosemide is administered @ 1.0 to 2.0 mg/kg every 1 to 2 hour until respiration rate decreases and comes to about 30-40 breath per minute. Constant rate infusion of furosemide is also a better option in acute cases. Significant hypokalemia may develop on furosemide therapy requiring constant monitoring of potassium level and supplementation of potassium in cats with lower potassium level.
- Torsemide, a potent diuretics, can replace furosemide. Though torsemide is often reserved for refractory cases, it may also be used as first line diuretic therapy. After reduction of respiration, furosemide may be given orally @ 1.0 to 2.0 mg/kg PO. BID
- Constant monitoring of renal function is essential in cats undergoing diuretic therapy.
- Stress and anxiety can be managed with intramuscular butorphanol given @ 0.1 to 0.2/kg body weight.
- Pleural effusions is more important cause of dyspnea in cats with heart failure and may require emergency thoracocentesis.
- Cats with heart failure have higher risk of aortic thromboembolism. Therapy with clopidogrel (18.75 mg /cat PO OD with food) should be started even as a prophylactic therapy. Retching, vomiting or salivation are some of the side effects of clopidogrel. These effects can be minimized by giving the drug in gelatin capsule followed by water. If thrombus development has already taken place, therapy with clopidogrel and aspirin should be initiated.
- Use of pimobendan in cats with heart failure is still inconclusive.
- Benazepril, ACE inhibitor, has shown no added advantage in cats with congestive heart failure (King *et al.*,2019).
- Cats showing significant arrhythmia should be treated with anti-arrhythmic drugs. Ventricular tachycardia is treated with lidocaine

(xylocaine) at a lower dose rate of 0.5 mg/kg intravenously as cats are sensitive to lidocaine. Supraventricular or atrial arrhythmias are to be treated with diltiazem (0.5-1.5 mg/kg PO TID with food).

- Cats with chronic heart failure are treated with oral administration of furosemide @0.5 -2.0 mg/kg PO BID/TID depending on severity with evaluation of renal function every 3 to 7 days. Benazepril does not delay the onset of treatment failure.
- Refractory cases of heart failure in cats can be managed with torsemide. Spironolactone @ 1.0 to 2.0 g/kg PO OD/BID is another drug of choice. In Maine Coon cats, ulcerative dermatitis has been observed on administration of spironolactone @ 2.0mg/kg PO BID
- Taurine supplementation (250 mg PO BID) is beneficial in cats with heart failure due to left ventricular systolic dysfunction.
- Reducing sodium chloride in the diet is advocated.
- Cases of aortic thromboembolism in cats require adequate analgesia during first 24 hours. Fentanyl, hydromorphine or methadone are the drugs of choice.
- Anticoagulant therapy with low molecular weight heparin should be instituted.
- Heart failure with aortic thromboembolism is treated with oxygen and furosemide. But tachypnea in aortic thromboembolism can be due to severe pain and needs to be differentiated from heart failure. Clopidogrel can be started with initial loading dose of 75 mg/cat orally followed by 18.75 mg PO once daily.
- Cats should be regularly evaluated at weekly interval for development of necrosis in distal limbs (if any), electrolyte status, kidney function, appetite, and degree of improvement in neuro-muscular functioning.

Treatment Protocol of Heart Failure as per Clinical Severity

a. Acute CHF

- Diuretics are the cornerstone of therapy for acute heart failure in felines also. Furosemide can be used as intermittent IV bolus (1-2 mg/kg IV every 30-60 minutes) until dyspnea is reduced and then it can be given twice or thrice daily. Alternatively furosemide can be administered as an initial IV loading bolus, followed by CRI (0.25-0.5 mg/kg/hr.).

- Transdermal application of 2% nitroglycerin ointment, ¼ to ½ inch every 6 hours enhances reduction of pulmonary edema. Then it may be used twice daily.
- Supplemental oxygen (40-60% 0_2-enriched inspired gas) may improve pulmonary gas exchange

b. Chronic CHF

- The aim of therapy in chronic congestive heart failure in cats is to maintain a congestion-free state, prevent arterial thromboembolism, slow the progression of myocardial dysfunction, promote enhanced quality of life, and prolong survival.
- Gradually decrease furosemide to the lowest effective dose level (6.25-12.5 mg OD/BID). Some cats may respond favorably at 1-2 mg/kg PO on alternate days or twice weekly
- In long term therapy, ACE inhibitor (enalapril, 0.5 mg/kg daily) may be combined with furosemide.
- Beta blocker may be added in cats with chronic heart failure due to dynamic LV outflow tract obstruction.

c. Recurrent CHF

- Diuretic resistance may occur as heart failure progresses.
- Dose of diuretics needs to be increased. Intravenous furosemide may yield good results
- Other diuretic (thiazide, 5-10 mg daily, or spironolactone, 12.5-25 mg daily) may be used in cats showing furosemide resistance..
- Enalapril can also be added.

Shock in Cats

Response of cats to shock is a little different. During compensatory stage of shock, heart rate is either normal or slow in cats. Low cardiac output results in to corresponding decrease in peripheral circulation, hypothermia. Hypothermia decreases arterial response to catecholamines and there is vasodilation rather than vasoconstriction. There is associated decrease in blood pressure. Hypothermia, bradycardia and hypotension is the hall mark of shock in cats. Shock in feline patients is clinically characterized by lethargy, depression, pale mucus membrane (Fig. 32), slow heart rate, weak/ absent pulse, low body temperature and low blood pressure.

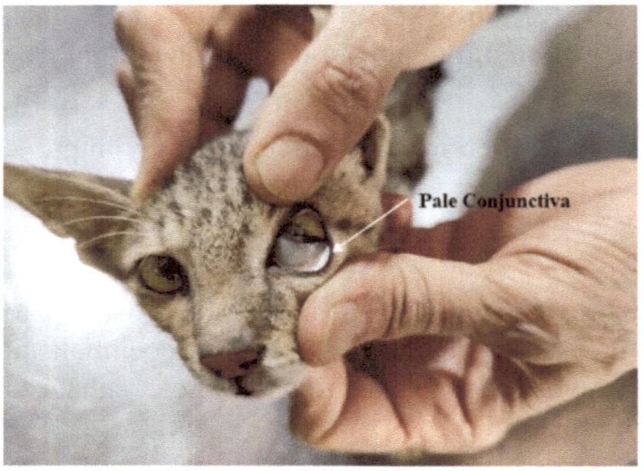

Fig. 32: A kitten (weighing 900 g , aged 4 month) showing pale conjunctiva and ear pinna, hypothermia (95.0F), bradycardia (heart rate 140 bpm), lethargy, depression, very weak femoral pulse , severe anemia (Hb 3.4 g/dl, PCV 9.0%), anorexia and increased capillary refill time (> 2 second) suggesting compensatory stage of hypoxic shock.

Abstract Clinical Findings in Different stages of shock in Cats

a. Compensatory Shock

- Rectal temperature normal to slightly low.
- Heart rate normal or slow.
- Pale mucus membrane.
- Hyperemic mucus membranes in distributive shock.
- Normal to slightly prolonged capillary refill time.
- Rapid capillary refill time in cases of distributive shock.
- Tachypnea.
- Normal to slightly low blood pressure .
- Mental alertness

b. Early Decompensatory Shock

- Rectal temperature slight to moderately low.
- Moderate tachycardia
- Pale or white mucus membranes.
- Prolonged capillary refill time.

- Tachypnea.
- Mild to moderately low blood pressure.
- Mental alertness reduced

c. Late Decompensatory Shock

- Markedly low rectal temperature (Fig.33,34, 35)
- Muddy/ gray/cyanotic mucus membranes (Fig. 33)
- Mild tachycardia or severe bradycardia (Fig.34 B, 35 B)
- Prolonged capillary refill time (Fig.33,34,35)
- Reduced respiration rate
- Marked hypotension unresponsive to fluid therapy
- Mental alertness markedly reduced.

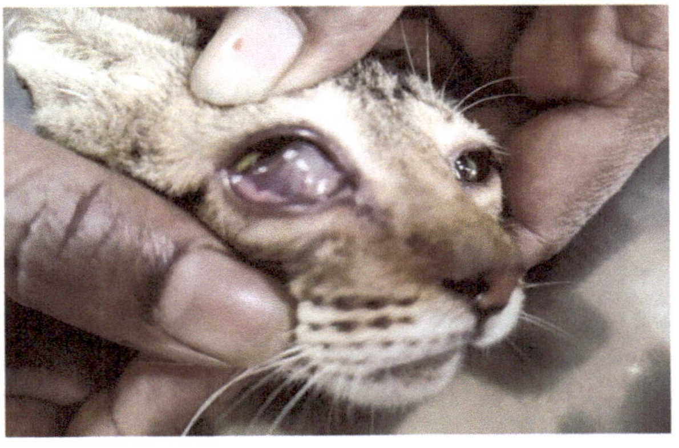

Fig. 33: A kitten (weighing 1100 g, aged 3.5 month) showing congested muddy conjunctiva, hypothermia (92.0 ^{0}F), severe bradycardia (heart rate 140 bpm), lethargy, depression, very weak femoral pulse, severe anaemia (Hb 3.0 g/dl, PCV 8.0%), anorexia, reduced respiration rate and increased capillary refill time (> 3 second) suggesting late decompensatory stage of hypoxic shock.

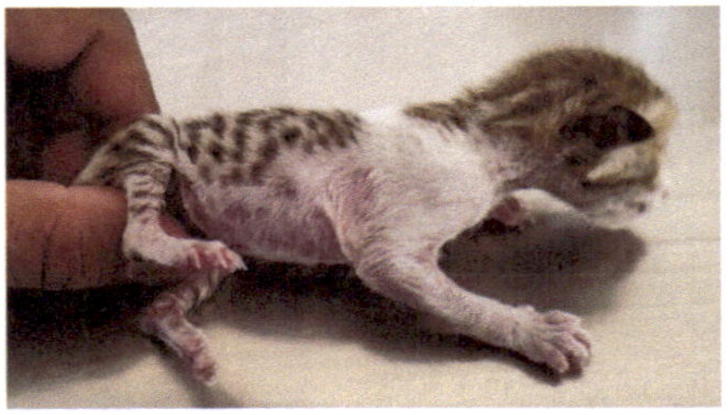

A

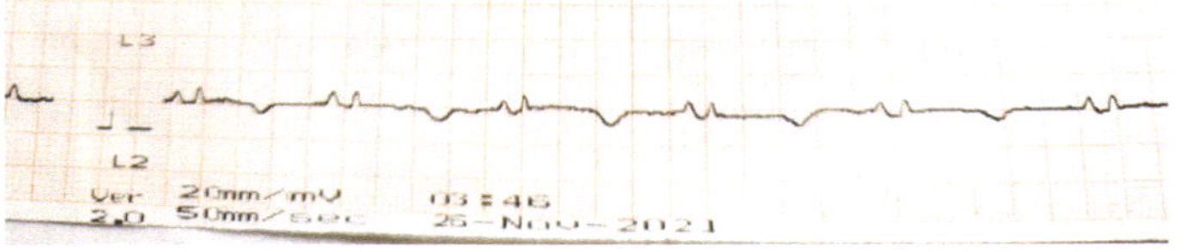

B

Fig. 34: An orphaned anorectic kitten (A) of one month age with hypothermia (< 90.0F), severe bradycardia (heart rate 125 bpm – B), lethargy, marked depression, imperceptible femoral pulse, anorexia, reduced respiration, increased capillary refill time (> 3 second), and severe anaemia (Total erythrocytes 2.3 million/mm^3, haemoglobin 3.0 g/dl, and packed cell volume 10.0%) and thrombocytopenia (platelets 29,000/mm^3) suggesting late decompensatory stage of shock

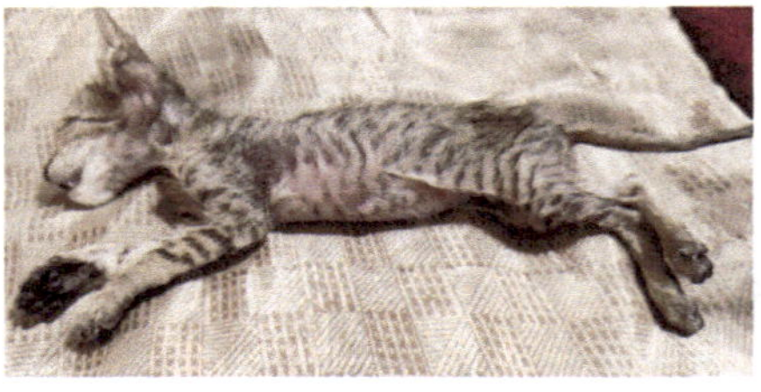

A

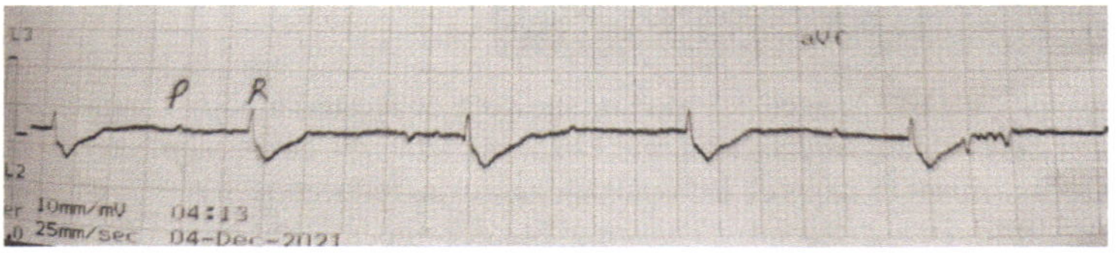

B

Fig. 35: An orphaned 5 days old recumbent kitten (A) showing hypothermia (< 90.0F), lethargy, marked depression, imperceptible femoral pulse , unopened eyes, reduced respiration (7 to 8 per minute) , no capillary refill, no mental alertness and no response to any stimuli. Electrocardiogram of the kitten (B) revealed bradycardia (HR 60 bpm);normal P (0.1 mV, 0.04 sec), R (0.3 mV), QRS (0.35 mV); prolonged P-R interval (3.8 sec to 0.56 sec): S-T segment slurring (0.2 mV): -ve T (0.15 mV, 0.07 sec) suggesting severe bradycardia, AV-block and myocardial hypoxia. On the basis of clinical picture and electrocardiogram, the kitten was diagnosed with late decompensatory stage of shock. All attempts to give intravenous fluid were futile as no vein could be raised for IV fluid transfusion owing to severe hypotension and the kitten collapsed.

Drugs used in Cardiac Diseases in Feline

Amlodipine @ 0.625-1.25 mg/cat PO OD

Atenolol @ 6.25 mg-12.5 mg /cat PO OD

Benezapril @ 0.25-0.50 mg/kg PO BID

Digoxin @ 0.0025- 0.004 mg /kg PO BID

Diltiazem @ 7.5 mg/cat PO OD

Hydralazine @ 1.0-3.0 mg/kg PO BID

Lidocaine@ 0.25 -0.5 mg/kg IV slow

Propranolol @ 0.02mg/kg slow IV . can be repeated 4 hrly.

Sotalol @2.0 mg/kg PO BID

References

Chetboul, V., Foulex, P.,Kartout,K.,Klein, A.M., Sailleau, C., Dumarest, M., Delaplace, M., Gouilh, M.A., Mortier, J. and Poder, S.L. (2021). Myocarditis and subclinical-like infection associated with SARS-CoV-2 in two catsliving in the same household in France: A case report with literature review. Front. Vet. Sci., 21 October 2021 | https://doi.org/10.3389/fvets.2021.748869

Goutal, C., Keir,I.Kenny,S. Rush,J. and Freeman,L. (2010). Evaluation of acute congestive heart failure in dogs and cats: 145 cases (2007-2008). J.Vet.Emergency Crt.Care 20:330-337.

King, J., Martin, M., Chetboul, V., Ferasin, L., French, A., Strehlau, G., Seewald, W., Smith, S., Swift, S., Roberts, S., Harvey, A., Little, C., Caney, S., Simpson, K., Sparkes, A., Mardell, E., Bomassi, E., Muller, C., Sauvage, J., Diquelou, A., Schneider, M., Brown,L., Clarke, D. and Rousselot, J. (2019). Evaluation of benezapril in cats with heart disease in a perspective, randomized, blinded placebo-controlled clinical trial. J.Vet.Intern.Med. 34 :1- Pub Date : 2019-09-29, DOI: 10.1111/jvim.15572

Novo Matos, J., Pereira, N., Glaus, T., Wilkie, L., Borgeat, K., Loureiro, J., Silva, J., Law, V., Kranjc, A., Connolly, D. and Luis Fuentes,V. (2017). Transient myocardial thickening in cats associated with heart failure. J.Vet.Intern.Med. 32:48-56.

Smith, S., Tobias, A., Fine, D., Jacob, K.A., and Ployngam, T. (2004). Corticosteroid-associated congestive heart failure in 12 cat. Intern. J. Appl. Res. Vet. Med. 2:159, 170.

11

Heart Failure in Ruminant

Heart failure is the terminal outcome of heart diseases. It is defined as ventricular dysfunction (pump failure) leading to cardiac inability to maintain an adequate cardiac output. Heart failure is the outcome of filling defect, cardiac muscle abnormality, conduction defect, an excessive work load or a combination of these abnormalities. Bovine congestive heart failure (BCHF) is being recognized as an emerging condition of well managed high genetic merit cattle (feedlot cattle) in the western Great Plains of the United States and Canada. BCHF is a fatal condition involving pulmonary hypertension leading to right ventricle failure. Unfortunately progressive compensatory mechanisms in neurohormonal axis in heart diseases have negative effects on the cardiac muscle and its functioning. Reduction in cardiac output results in bankruptcy of oxygen and nutrients in tissues and failure to maintain circulatory stability'. In ruminants' predominant clinical signs of heart failure are related to right heart. While left heart failure is more common in dogs, cats and horses. Circulatory failure is manifested as congestive heart failure, acute heart failure (cardiac syncope) and peripheral circulatory failure.

Etiology

Heart Failure in ruminants may be due to following diseases.

- Bacterial endocarditis.
- Traumatic reticulopericarditis.
- Lymphoma involving cardiac muscles.
- Cardiomyopathy .Dilated cardiomyopathy has been reported in Holstein (Nart *et al.*, 2004) and Ayrshire cattle (Martig and Tschudi ,1985).
- Secondary cardiomyopathies have been observed in ruminants due to intoxication of monensin (Van Vleet *et al.,* 1983), deficiency of vitamin E or selenium and infections (bacterial or viral).
- Ventricular Septal defects.
- Congenital heart diseases.

- Genetic etiology is being investigated in beef feed lot cattle with bovine congestive heart failure.

Clinical Manifestations

Ruminants suffering from heart failure may have a variety of clinical symptoms as given below.

- Most of the signs of heart failure in ruminants are related to right heart failure.
- Peripheral edema.
- Lung edema.
- Brisket edema (Fig. 36) / submandibular edema.

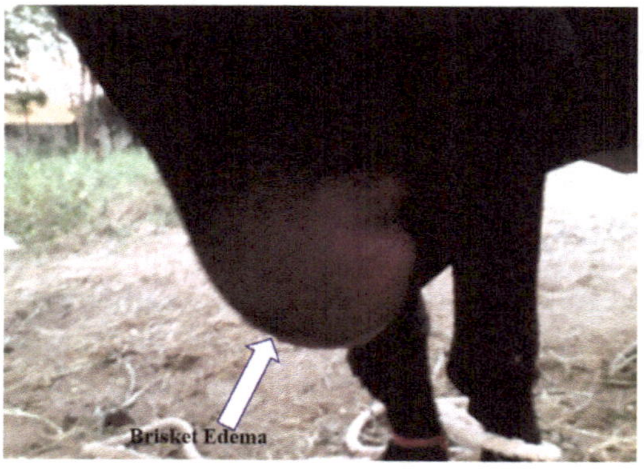

Fig. 36: Brisket edema in a buffalo due to traumatic reticulo-pericarditis.

- Abducted elbow (Fig.37)

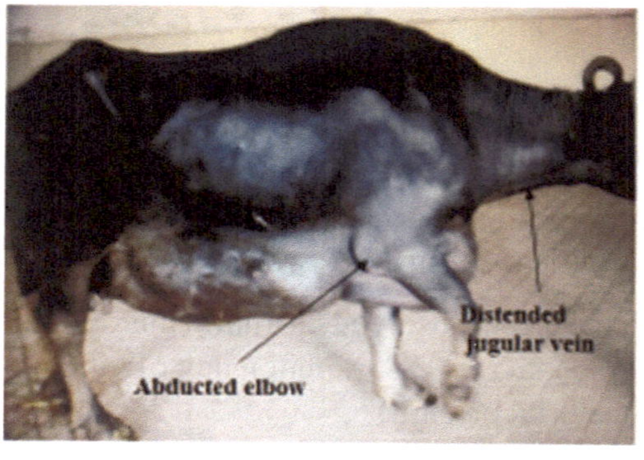

Fig. 37: Abducted elbow and distended jugular vein in a buffalo with traumatic reticulo-pericarditis.

- Jugular vein distension or pulsation (Fig.37)
- Weakness.
- Exercise intolerance.
- Abnormal respiration (tachypnea).
- Abnormal respiratory sounds (crackle, wheezes).
- Hepatomegaly
- Cough very rarely.
- Hypo motile rumen.
- Dehydration.
- Anorexia

Diagnosis

- Clinical signs such as brisket and peripheral edemas, jugular distension, tachypnea, exercise intolerance are suggestive signs of right heart failure.
- Electrocardiographic Changes- Following arrhythmias may be seen in ruminants with heart failure.

  Tachycardia.

  Atrial Fibrillation.

  Ventricular premature complex
- Echocardiographic Changes- Echocardiography may reveal following pericardial and pleural changes in traumatic reticulo-pericarditis cases.

  Pericardial effusions.

  Thickening of pericardium.

  Pleural effusions.
- X-ray Examination- Radiography may be helpful in detecting piercing metallic objects in cases of traumatic reticulo-pericarditis.

  Detection of penetrating metal objects into pericardium (Fig. 38).

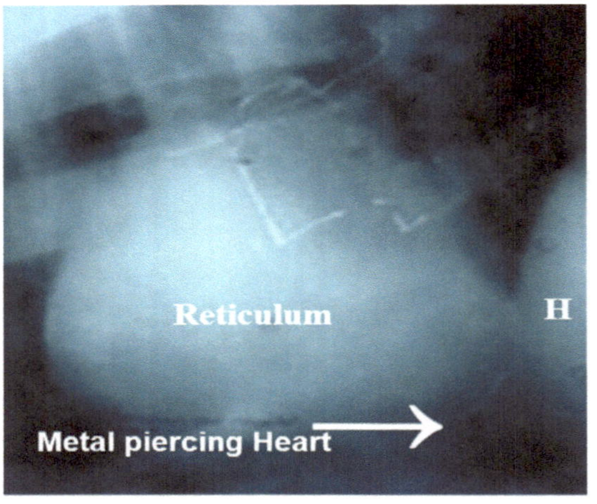

Fig. 38: Radiograph of a buffalo showing piercing metallic material in the heart.

- Elevation in cardiac biomarkers- Laboratory investigations may reveal higher values of SGOT and cTn-I.

  Increase in SGOT level (Prasad and Joshi,, 1973)

  Increase in cTn-I level (Sarita Devi *et al.,*2014)

Treatment Protocol

- Furosemide is the main drug to control edemas, jugular distension and to reduce cardiac preload.
- Correction of electrolytes is beneficial.
- Periodic electrolyte monitoring to avoid hypokalemia, hyponatremia and hypocalcaemia that may develop on diuretic therapy.
- Use of positive inotropes is considered valuable but needs to be used cautiously because of its side effects. Positive inotropes should not be used in heart failure due to monensin intoxication.
- Serious arrhythmias are treated with specific antiarrhythmic drugs.
- Antibiotics are employed to control infection.
- Pericardiocentesis should be done in cases of pericardial effusion associated with traumatic reticulopericarditis.

Prognosis

- Prognosis of cases with heart failure in large ruminants is poor.

Peripheral Circulatory Failure

Failure of venous return to heart results into reduced cardiac output leading to peripheral circulatory failure. Decreased flow causes tissue hypoxia leading to decreased tissue function. Peripheral circulatory failure occurs due to decrease in circulatory blood volume leading to decreased venous return. Heart functioning is adequate in peripheral circulatory failure.

Etiology

- Precipitating causes of peripheral circulatory failure are hypovolemic failure, distributive failure, nontoxic general vasogenic failure, toxic and septic shock.
- Hypovolemic failure can occur in cases of severe hemorrhage, dehydration (colitis-x, severe stress, neonatal calf diarrhea), and fluid loss into intestines (acute intestinal obstruction, toxic shock).
- Distributive failure is observed in vasogenic shock or vascular abnormalities generally associated with severe extensive burns, following extensive surgery, uterine prolapse, rapid abdominocentesis to reduce ascites, severe colicky pain and trauma with local sequestration of blood and fluid.
- Non-toxic general vasogenic failure are seen in parturient paresis in dairy cows and in hypocalcaemia in sheep.
- Toxic and septic shocks precipitate peripheral circulatory failure. Endotoxins are the cause of toxic shock. Infectious acute- diffuse peritonitis, acute gangrenous mastitis, per acute coliform mastitis and acute septic metritis may lead to septic shock.

Patho-physiology

- Falling cardiac output due to hypovolemia stimulates sympathetic nerves and release of catecholamine from adrenals causing vasoconstriction, fall in blood pressure, decreased perfusion of the organs, hypoxia and tissue acidosis. Decreased perfusion of kidneys causes oliguria, metabolic acidosis. Continued reduction in cardiac output causes damage to myocardium , nervous system and other organs, disseminated intravascular coagulopathy; adversely affects vascular reactivity and permeability (due to release of archidonic acid that lands up in the production and release of prostaglandins, thromboxane and leukotriene) leading to shock.

Clinical manifestations

- Muscular weakness
- Reduction in body temperature
- Fall in blood pressure
- Pale mucus membrane
- Increased capillary refill time
- Increased heart rate
- Small weak pulse
- Increased respiration rate
- Anorexia
- Thirst
- Depression
- Listlessness
- Coma in terminal stage
- Mild clonic convulsions in few cases.

Diagnostic Features

- There is no specific diagnostic tests. It is by inclusion and exclusion of various factors.
- Haeogram may reveal haemoconcentration in animals with dehydration or shock,
- Evidence of circulatory failure with no detectable abnormality of the heart and the presence of primary etiology such as extreme dehydration, hemorrhage is indicative of peripheral circulatory failure.
- Blood pressure monitoring can be an important tool of diagnostic assistance.
- Alteration in acid-base status and metabolic acidosis may be seen in inadequately treated animals

Treatment Guidelines

- To restore the circulatory blood volume is the main goal of therapy.
- Intensive monitoring and fluid replacement is needed in animals with hematogenic failure having packed cell volume <20 %. Blood transfusion is needed in animals with haematogenic failure having packed cell volume <12 %.
- Animals having peripheral circulatory failure showing total serum proteins < 3.5 g/dl should be treated with plasma or whole blood transfusion.
- To restore circulatory blood volume, large amount of fluid transfusion is needed. Measuring of central venous pressure can be used to guide the amount of fluid to be administered. Measuring central venous pressure is generally not possible in routine clinical practice. In this circumstance, clinical indicators like jugular vein pressure, quality of arterial pulse, restoration of urine flow are of clinical assistance in assessing the quantity of the fluid to be administered.
- Bicarbonate therapy should be initiated in animals with marked lactic acidosis.
- There is no justification of inotropes in cases of peripheral circulatory failure as there is no primary abnormality of the heart.
- Flunixin meglumine (cyclooxygenase inhibitor) can have additional therapeutic advantage if given early in cases of endotoxic shock, toxic shock and toxemia.

Acute Heart Failure

Acute heart failure in ruminants is clinically characterized by sudden unconsciousness, falling with or without convulsions, pale mucus membrane, collapse or recovery. Acute heart failure should always be considered as one of the possibility in ruminants collapsing suddenly after exertion or excitement.

Etiology

Acute heart failure in ruminants may occur in following conditions.

- Severe tachycardia (associated with FMD virus, encephalomyocarditis virus; deficiency of copper or selenium; electrocution/ electric strike ; Phalaris spp. plant toxicity)

- Severe bradycardia (associated with intravenous calcium, xylazine, potassium chloride concentrated solution)
- Sudden increase in cardiac work load (anaphylaxis, rupture of aortic valve)
- Tachyarrhythmias associated with severe excitement (due to increase catecholamine).
- Pericardial tamponade, rupture of aortic and pulmonary artery.
- Drugs affecting cardiac conduction may also lead to acute heart failure in the absence of primary heart disease,
- Ingestion of certain toxic plants.

Pathogenesis

- Arrhythmias (severe tachyarrhythmias or bradyarrhythmias) associated with acute heart failure result into ventricular filling defect finally leading to reduced cardiac output. That in turn causes fall in minute volume of the heart leading to severe degree of tissue anoxia. Brain is the most sensitive to anoxic state and is adversely affected resulting into nervous symptoms. Reduction in arterial blood flow leads to pale mucus membranes and weakness. Respiratory distress is the predominant manifestation in less severe cases of acute heart failure owing to pulmonary edema. These cases are better described as cases of acute congestive heart failure.

Clinical manifestations

Acute heart failure may occur in ruminants when they are at rest, excited or doing activity. Animals with acute heart failure may show a variety of clinical signs as follows.

- Dyspnea, staggering, falling and sudden collapse.
- Pale mucus membrane
- Limb incoordination.
- Arrhythmias (bradycardia or tachycardia).
- No palpable pulse or no heart sound.
- Severe tachyarrhythmia such as multiple ventricular extra systole in enzootic muscular dystrophy in calves.
- Respiratory distress in less severe cases owing to pulmonary edema.

Diagnosis

- In acute heart failure there is hardly any time for laboratory investigations. Nevertheless, elevated levels of cardiac biomarkers (troponins, BNP) are suggestive of myocardial disease.
- There is no specific test to diagnose acute heart failure. However, severe arrhythmias (electrocardiographic investigation), pale mucus membrane, absence of palpable pulse and mild convulsions may be suggestive of acute heart failure.
- On necropsy, engorgement of visceral veins may be observed if episode of heart failure was of a few minute.

Treatment Approach

- Because of short course, undertaking treatment becomes impracticable.
- Cardiac massage or electrical stimulation can be undertaken in cases of sudden cardiac arrest or ventricular fibrillation during anaesthesia.

Congestive Heart failure

Congestive heart failure is characterized by inability of heart, due to intrinsic factors, to maintain circulatory equilibrium at rest resulting into venous congestion, edema/ congestion of lungs and /or periphery, ascites, heart enlargement and arrhythmias.

Etiology

Diseases of the heart is the main cause of congestive heart failure in ruminants. Etiological factors responsible for congestive heart failure are enumerated below.

- Cardiac muscle diseases

  Infectious myocarditis (viral, bacterial, parasitic)

  Degenerative myocardial diseases (toxic or nutritional deficiencies)

  Hereditary or congenital myocardial disease

  Cardiac neoplasm (bovine viral leucosis)
- Conduction disturbances
- Valvular defects

  Valvular stenosis

Valvular insufficiency (aortic valve insufficiency, mitral valve insufficiency).

Valvular rupture

- Endocarditis
- Pericardial diseases

Pericarditis

Pericardial tamponade

Hydropericardium

- High altitude disease
- Congenital septal defects
- Congenital vascular abnormalities.

Pathogenesis

- With reduced contractility and increased load in cardiac diseases compensatory mechanism starts operating with increase in heart rate, increase ventricular filling, blood flow redistribution, dilatation and hypertrophy to maintain circulatory equilibrium.
- Nevertheless, cardiac reserve is low and animal is unable to cope up with the increased demand of exercise or physical activity leading to exercise intolerance.
- When these efforts reach their physiological limits and fail to maintain circulatory requirements of the body at rest, congestive heart failure develops.
- Congestive heart failure may be due to left heart failure or right heart failure or both ventricular failure.
- Right heart failure is associated with venous congestion in greater circulation leading to compromised functioning of liver and kidneys.
- Poor perfusion of kidneys leads to reduced urine output.
- Damage to glomeruli, due to poor perfusion, causes leakage of plasma proteins into the urine.
- Venous congestion in portal system, a sequel of hepatic congestion, leads to impaired digestion , absorption and more transudate in intestines resulting into diarrhea.

- Pulmonary congestion and edema is conspicuous in left heart failure with sudden onset of volume overload on the left ventricle.

Clinical manifestations

- Evidence of cardiac enlargement, moderately increased resting heart rate and respiratory distress on light exertion may be observed in early stage of heart failure.
- Congestive left heart failure is associated with an increased rate and depth of the respiration at rest, moist crackles at the base of lungs and an increased dullness at ventral borders of the lungs on percussion. Dyspnea and cyanosis is seen in terminal stage.
- Congestive right heart failure is associated with increased heart rate, ascites, anasarca (edema of neck, jaw and ventral surface of body), hydrothorax , hydropericardium, enlarged palpable liver (protruding beyond right costal arch), small volume of urine with small amount of albumin, jugular vein distension (due to an increase in end diastolic pressure of right ventricle).

Diagnosis

- There is no definite diagnostic test to confirm congestive heart failure. Manifestations described in congestive heart failure are also seen in other diseases.
- Therefore a comprehensive approach starting from detailed history, clinical symptoms, radiographic examination, electrocardiographic examination, echocardiographic examination, to evaluation of cardiac biomarkers may be of diagnostic assistance.

Prognosis

- Though, prognosis of congestive heart failure varies with the cause, it is largely grave in large ruminants.

Treatment protocol

- Improving myocardial contractility, reducing preload and reducing the demand on cardiac output are the main stay of therapy as in other animals.
- Myocardial contractility can be improved by using digoxin, commonly used cardiac glycoside. In ruminants it is used intravenously as it is

destroyed in the rumen when given orally. Intramuscular administration of digoxin is contraindicated as it causes severe muscular necrosis. Use of digoxin increases myocardial contractility, improves cardiac output, reduces heart rate, decreases heart size, and increases myocardial oxygen consumption. Improvement in cardiac output and peripheral circulation promotes diuresis and thus reduces edema. In ruminants digoxin has been recommended at a loading dose of 0.022 mg/kg followed by 0.0034 mg/ kg every 4 hours or @ 0.086 mg/100 kg as a constant rate infusion. Use of digoxin may be associated with toxicity in some animals with congestive heart failure.

- For reducing preloads and edemas, diuretics are the drug of choice. Furosemide is commonly used for this purpose @ 2.5 to 5.0 mg/kg. Acetazolamide and chlorothiazide are other diuretics.
- Free access to salt should be prohibited.
- Stall rest should be considered to reduce the circulatory demand on heart.
- Anorectic cattle on furosemide therapy should be supplemented with potassium chloride (cattle 100 g daily orally) with regular monitoring of blood potassium level.

Heart Failure in Goats

Heart diseases are rarely diagnosed in goats.

Etiology of Heart Failure

- Nutritional myopathy (Ross *et al.*,1989)
- Pericarditis (Akkoc, 2007)
- Bacterial endocarditis
- Congenital heart defects (Gardner *et al.*, 1992; Laus *et al.*,2011)
- Intoxication (East *et al.*, 1994)
- Neoplasms (Braun *et al.*, 1995),
- Dilated cardiomyopathy with mitral and tricuspid insufficiency (Varshney *et al.*, 2022).
- Thymoma. Congestive heart failure in two adult Nubian goats had been reported in association with thymoma in anterior mediastinum (Rostkowski *et al.*, 1985)..

Clinical Signs

As in other species heart failure is the end stage manifestation of heart diseases. The clinical signs of heart failure in small ruminants and cattle are similar

- Jugular vein distension and pulsation (Fig.39)

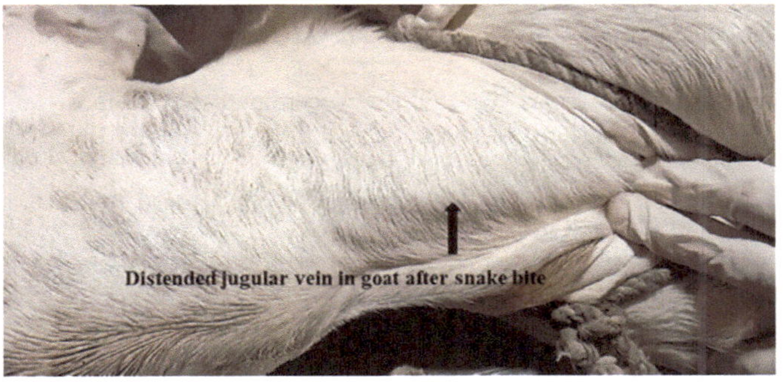

Fig. 39: Showing distended jugular vein in a collapsing goat 12 hour after snake bite.

- Peripheral edema
- Ascites
- Respiratory distress/tachypnea
- Coughing
- Pale mucus membrane
- Tachycardia

These signs are the main indicators of right-sided heart failure (Pugh, 2002).

Diagnosis

- The ancillary tests for diagnosing heart diseases are the same in small ruminants and in cattle.
- The echocardiography is of great assistance in assessing heart dimensions (Olsson *et al.*, 2001), changes in myocardium, valvular insufficiency (Gardner *et al.*, 1992) and the pericardium (Braun *et al.*, 1995) in small ruminants.
- When the diagnosis of congestive heart failure is made, the prognosis is grave to poor in small ruminants (Pugh, 2002).

Treatment

- Therapeutic approach of cor pulmonale is limited.
- Treatment is similar as described in other ruminants.

References

Akkoc, A. (2007).Traumatic reticulopericarditis in a Saanen goat. Turk. J. Vet. Anim. Sci. 31: 283 – 285.

Braun, U., Hagen, A., Pusterla, N. and Caplazi P. (1995). Echocardiographic diagnosis of a cardiac fibrosarcoma in the right atrium of a sheep. Schweiz. Arch. Tierheilk. 1995, 137: 187 – 192.

East, N.E., Anderson ,M., Lowenstine, L.J. (1994). Apparent gossypol induced toxicosis in adult dairy goats. J. Am. Vet. Med. Assoc. 204: 642 – 643.

Gardner, S.Y., Reef, V.B., Palmer, J.E., Reimer, J.M., Sweeney, R.W.(1992). Echocardiographic diagnosis of an anomaly of the tricuspid valve in a male dairy goat. J. Am. Vet. Med. Assoc. 200:521-523.

Laus, F.,Copponi,I., Cerquetella,M. and Fruganti,A. (2011). Congenital cardiac defects in a pygmy goat (Capra hircus). Turk J. Anim .Sci.35:471-475.

Martig, J. and Tschudi, C. (1985).Weitere Fälle von Kardiomyopathie beim Rind. *Dtsch Tierarztl Wochenschr.* 1985;92:363–366. [PubMed] [Google Scholar]

Milne, M.H., Mellor, D.J, Barrett, D.C. and Fitzpatrick,J.L. (2001). Observations on ascites in nine cattle. *Vet Rec.* 2001;148:341–344. [PubMed] [Google Scholar]

Nart, P., Thompson, H., Barrett, D.C., Armstrong, S.C.McPhaden, A.R. (2004). Clinical and pathological features of dilated cardiomyopathyin Holstein-Friesian cattle. Vet. Rec.155:355-361.doi:10.1136/vr155.12.355

Olsson, K. ,Hansson, A., Hydbring, E., Von Walter, L.W. and Häggstrom, J.(2001). A serial study of heart function during pregnancy, lactation and the dry period in dairy goats using echocardiography. Exp. Physiol. 6: 93 – 99.

Prasad, J. and Joshi, B.P. (1973). Some biochemical studies in traumatic pericarditis in a buffalo. Indian Vet. J. 50: 148-151.

Pugh, D.G. (2002). Sheep and Goat Medicine. W. B. Saunders, Philadelphia.pp 1 – 17.

Ross, A.D., Gee, G.C., Jackson, A.R., Hall, E. and Greentree P. L. (1989). Nutritional myopathy and goats. Aust. Vet. J. 66: 361 – 363.

Rostkowski, C.M., Stirtzinger, T., and Baird, J.D. (1985).Congestive heart failure associated with thymoma in two Nubian goats. Can.Vet.J.26:267-269.

Sarita Devi, Varshney,J.P.,Undirwade, S.C. and Jadav, K.M.(2014). Cardiac troponin-I and electrocardiographic findings in buffaloes with traumatic pericarditis. Ruminants Sci. 3:123-126.

Van Vleet, J.F., Amstutz, H.E., Weirlich, W. E., Rebar, A.H. and Ferrans,V.J. (1983). Clinical, clinicopathologic, and pathologic alterations in acute monensin toxicosis in cattle. *Am J Vet Res.* 1983;44:2133–2144. [PubMed] [Google Sc holar]

Varshney, J.P., Chaudhary, P.S. and Saini, N. (2022).Sonograms in Different Diseases of Animals: Case Studies. In: Ultrasound in Veterinary Medicine: Fundamentals and Application. Varshney,J.P. (ed.), NIPA, a division of NIPAGenx ElectronicResources and Soluttions P. Ltd., New delhi-110034. p361.

12

Heart Failure in Equine

Heart failure is not a disease but it is the end stage of heart diseases. Heart failure in horses is not very common as in dogs and cats. In horses it is characterized by the signs of forward failure, backward failure and fall in blood pressure. Actually heart failure is a complex process involving not only structural abnormalities and hemodynamic mechanism but also neuroendocrine, biochemical and genetic pathways resulting into either an increased hemodynamic burden or reduction in myocardial oxygen delivery leading to abnormal myocardial structure and function. Information regarding prevalence of congestive heart failure in horses is insufficient and may vary in different horse populations in different places.

Etiology

There are many conditions that may lead to heart failure in horses.

- Left Heart or biventricular heart failure is caused by mitral valve insufficiency (rupture of a chorda tendinae , degenerative mitral valve disease, or infective endocarditis), aortic insufficiency (rupture of an aortic valve leaflet, infectious endocarditis), myocardial disease (myocarditis, myocardial fibrosis, myocardial ischaemia, dilated cardiomyopathy), and congenital heart diseases (ventricular septal defect, patent ductus arteriosus, aortocardiac fistula). Dilated cardiomyopathy has been associated with left heart failure in horses.
- Right heart failure is generally caused by pericardial effusions, pericarditis (effusive, constructive), tricuspid insufficiency (rupture of chorda tendinae, infective endocarditis), and congenital diseases (atresia of right atrioventricular valve, Tetralogy of Fallot, atrial septal defect, anomalies of great vessel) .

Pathogenesis

- When any significant heart disease occurs, compensatory mechanism sets in to maintain circulatory balance. Blood vessels constriction and increase in heart rate begins for a short period. Later on hormonal compensatory mechanisms start playing their role to increase cardiac output, blood pressure, blood volume and blood flow to compensate various degree of changes occurring due to heart disease. Finally long term activation of body compensatory response to progressive heart failure further complicates the disease pathogenesis and accentuate heart failure.

Clinical Manifestations

The following clinical signs may be observed in horses suffering from heart failure. Signs are generally attributed to reduced cardiac output and increased ventricular filling pressure.

- Increased heart rate
- Murmurs on cardiac auscultation
- Distended jugular veins and jugular pulsation
- Increased respiration
- Decreased appetite or anorexia
- Weakness
- Weight loss
- Edema of ventral abdomen and lower limbs
- Exercise intolerance
- Restlessness
- Pulmonary edema in left heart failure, mitral valve insufficiency and large ventricular septal defect. Signs of left heart failure generally go unnoticed with slowly progressive disease until right heart failure develops due to pulmonary hypertension.
- Coughing
- Absence of respiratory sounds over ventral lung field is suggestive of pleural effusions that frequently develops with heart failure.
- Advanced cases of heart failure may be associated with dyspnea, coughing and profuse white or pink tinged frothy nasal discharge
- Reduced renal output
- Collapse.

Classification of Heart Failure

Heart failure in horses is classified into following four functional categories.

- Systolic Myocardial failure -This type of cardiac failure is associated with reduction in myocardial contractility. Reduced contractility of the heart results in forward and backward failure. Echocardiographically there is reduced wall motion during systole. This may occur in many conditions such as trauma, infections, heat stroke, tumors, electric shock, poisons and drugs intoxication.
- Pressure Overload - Pressure on ventricular wall in increased
- Volume overload - Volume of blood in the chamber is increased. This type of heart failure occurs in valvular insufficiencies, and left to right shunts.
- Impedance to cardiac inflow - This type of heart failure may occur when blood flow is obstructed due to compression of the heart. There is a decrease in blood flow into and out of the heart. It may happen in pericardial effusions,

Diagnosis

- Clinical signs enumerated above are also seen in disorders of different body systems. Therefore signs of heart failure are not very specific. Peripheral edemas are also seen in conditions with protein loss (protein loosing enteropathy), in conditions with obstruction to lymphatic drainage (lymphoma, lymphadenopathy), and peripheral vasculitis (purpura haemorrhagica). Mild to moderate respiratory distress is also seen in respiratory diseases. Even acute pulmonary distress is seen not only in left heart failure but also in interstitial pneumonia and acute lung injury. Signs of low cardiac output are also nonspecific.
- Atrial fibrillations are common electrocardiographic findings in cases of congestive heart failure in horses (Prutton and Magdesian, 2016).
- Increased heart rate (>60 bpm), loud cardiac murmurs and echocardiographic detection of any valvular or cardiac defects in horses with atrial fibrillation is suggestive of congestive heart failure.
- Echocardiography is most valuable modality in the diagnosis of horses with heart failure.
- Measurement of pulmonary artery and capillary wedge pressure can document pulmonary hypertension.

- Radiography is valuable tool for identifying pulmonary edema as in other species also.
- Electrocardiography can assists in identifying arrhythmias and conduction disturbances.

Prognosis

- Long term prognosis of horses with heart failure is unfavourable

Therapeutic Protocol

- Poor contractility of ventricular muscle, arrhythmias, increased/ decreased blood pressure, reduced blood flow and decreased ventricular filling before systole occurring in heart failure due to cardiac structural changes further damage the heart and blood vessels if cases are left untreated. Therefore for the quality of life of the animal, addressing these problem is of great significance. Diuretics, positive inotrope and ACE-inhibitors are the drugs of choice to manage heart failure. Though, intravenous pimobendan has shown positive inotropic and chronotropic effects in healthy horses, its use in the management of heart failure in horses warrants further investigations (Afonso *et al*.,2016).
- Diuretics are the main stay of the heart failure therapy to reduce the congestions and fluid accumulation and to reduce cardiac over load. Furosemide @ 0.5 mg/kg IM once daily has been advocated. Torsemide, another loop diuretics, has also been used @ 4.0 mg/kg/day orally in cases of heart failure in horses.
- ACE-inhibitors are used to dilate the blood vessels and to lower down blood pressure .Out of four ACE-inhibitors (benezapril @ 0.25 mg/kg and 0.5 mg/kg; ramipril @ 0.1 mg/kg and 0.3 mg/kg; quinapril @ 0.125 mg/kg and 0.25 mg/kg; and perindopril @ 0.05 mg/kg and 0.1 mg/kg administered intragastically), oral administration of benezapril @ 0.5 mg/kg has been found effective in healthy horses (Afonso *et a*l.,2013). Oral administration of enalapril @ 0.5 mg/kg in horses with mitral valve insufficiency has shown an effect on the renin-angiotensin-aldosterone system (Gehlen and Rapp, 2020).
- Digoxin is the positive inotropic and negative chronotropic drug. It is recommended to increase the contractility of heart muscle. It is used @ 0.0022 mg/kg IV or 0.011 mg/kg PO BID
- Beta blockers and calcium channel blockers are used to reduce heart rate and to control atrial arrhythmias particularly atrial fibrillations.

- Persisting ventricular arrhythmias may be treated with intravenous lidocaine (20-50 microgram/kg/min as constant rate infusion).

References

Afonso,T., Giguere,S., Rapoport,G., Berghaus,L.J., Barton,M.H. and Coleman, A.E. (2013). Pharmacodynamic evaluation of 4 angiotensin converting enzyme inhibitors in healthy adult horse. J.Vet.Intern. Med. 27: 1185-1192.

Afonso,T.,Giguere, S., Rapoport,G., Barton,M.H., and Coleman, A.E. (2016). Cardiovascular effet of pimobendan in healthy mature horses. Equine Vet.J. 48: 352-356.

Gehlan,H. and Rapp,M. (2020). Effect of ACE inhibitor enalapril on plasma ACE concentration in horses with left side heart valve insufficiency. Equine Medicine 36 137-142.*DOI:* 10.21836/PEM20200206

Prutton, J.S.W. and Magdesian,K.G.(2016). Congestive heart failure in 51 Equics (1995- 2015). Equine Vet.J. 48: 6-6. https://doi.org/10.1111/evj.01_12612.

13

Heart Failure in Leporine

With an increase in domestication of rabbits, a leporine, as pet in urban dwellings and their increased life span, risk of cardiac diseases has also increased. Heart diseases are being diagnosed in rabbits with an increasing frequency. The development of clinical signs of heart failure is gradual in rabbits causing late recognition of the heart disease. Congestive heart failure in rabbits is a clinical condition in which lung edema, pleural effusion, hepatomegaly or all three pathologies may develop as a result of cardiac dysfunction. As a matter of fact any rabbit with severe respiratory distress having no upper respiratory sign should be suspected and evaluated for heart failure.

- Diseases of heart are more prevalent in rabbits of more than 4 year age. However, cardiac diseases are also seen in young rabbits.
- Heart failure can be observed in rabbits irrespective of breed, age or sex.
- Giant breeds of rabbits (French lop, Rexes, or New Zealand White) are considered more susceptible.
- The size of the thorax in rabbits is small as compared to abdomen and an increase in heart size in small thorax leads to respiratory distress in rabbits with heart failure.

Etiology

- Heart failure in rabbits is due to heart diseases (cardiomyopathy, vulvular insufficiencies, congenital heart defect, arrhythmias, severe anemia, or infectious diseases etc.). Viruses (Corona virus), bacteria (*Clostridium piliformis*, *E. coli*, *Pasteurella multocida*), or protozoa (*E. cuniculi*) have also been reported to cause cardiomyopathy leading to heart failure in rabbits . Stress and administration of certain drugs (doxorubicin, repeated anaesthesia with ketamine-xylazine, detomidine) may adversely affect the heart eventually leading to heart failure.
- Diets rich in fat along with lack of exercise is generally considered as an etiological factor for heart diseases in rabbits. But cardiac diseases have

also been recorded in active rabbits.

- Deficiency of vitamin E can adversely affect the myocardium in rabbits and may lead to heart failure.
- Stress induced cardiomyopathy has also been reported in rabbits. Cardiomyopathy may lead to heart failure
- Hypervitaminosis-D (due to errors in feed formulations) may cause calcification of the walls of the blood vessels owing to excessive absorption of calcium leading to heart failure.
- Rabbits in high altitude may have pulmonary hypertension leading to heart failure.
- Though hypertrophic cardiomyopathy and dilated cardiomyopathy have been reported in domestic rabbits (Houston *et al.*,2012), dilated cardiomyopathy is more common in rabbits.

Clinical Manifestations

- Cardiac diseases generally remain asymptomatic in early stage.
- Signs of heart failure in rabbits develop slowly and varies from decreased appetite, anorexia, fatigue, exercise intolerance, dyspnea, panting, ascites, edematous extremities (Vennen and Mitchell, 2009), hepato and splenomegaly. At advanced stage of heart failure, cyanotic mucus membrane, severe dyspnea, dilated nostrils, ascites, syncope, collapse and death may be evident. Coughing in rabbits with heart failure is not a common sign.

Diagnosis of Heart Failure

- Diagnosis of heart failure is assisted by clinical, radiographic, electrocardiographic and echocardiographic examinations.
- Murmurs can be detected on heart auscultation in rabbits with heart failure due to valvular insufficiency.
- Radiographic examination is the best examination to reveal pulmonary edema (Fig. 40), pleural effusions, heart enlargement in cases of congestive heart failure.
- Pericardial effusions, peritoneal effusions (ascites), peripheral edema, and hepatomegaly can be detected by radiographic examination in cases of right side heart failure.
- Electrocardiography can assist in detecting arrhythmias and tachycardias.

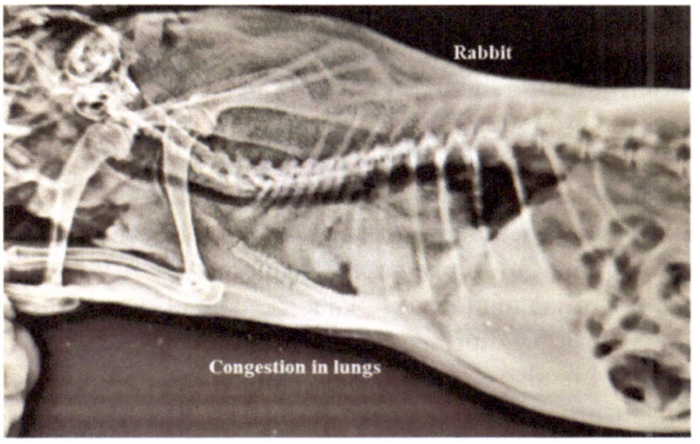

Fig. 40: Right lateral chest radiograph of a four year old male New Zealand White rabbit with dyspnea and tachycardia (heart rate 280 bpm) showing lung edema.

- Valvular insufficiency, cardiomyopathy, septal defects, pericardial effusions can be pinpointed by echocardiography.
- Cardiac troponin-I may be a predictive indicator of an early cardiac damage in rabbits also.

Treatment of Heart Failure

- The aim of treatment in heart failure in rabbits is to improve cardiac function (contractility, cardiac output), reduce pulmonary edema and stress. Use of diuretics reduces pulmonary edema and peritoneal effusions. Severe dyspnea owing to pleural effusions require pleurocentesis. Arrhythmias can be controlled with the use of anti-arrhythmic drugs.
- Emergency treatment in cases of acute heart failure consists of oxygen therapy (via oxygen tent, face mask), minimizing stress, thoracocentesis (to reduce pleural effusions if any), diuretics (furosemide (@ 1.0-2.0 mg/kg intravenously or intramuscularly), and the use of vasodilators (nitro glycerine trinitrate), ACE-inhibitors (enalapril 0.25 -0.50 mg/kg PO OD) and pimobendan (0.1-0.3 mg/kg OD).
- In acute cases prognosis is not very good as cases are generally presented in advance stage.
- Treatment in cases of chronic heart failure consists of diuretics (furesomide orally), ACE- inhibitors (enalapril or benazepril orally) and positive inotrope (digoxin orally). Furosemide is still effective diuretic in rabbits despite their relative inability to excrete H+ ions through

the kidney. As in dogs and cats enalapril inhibits the conversion of angiotensin I to angiotensin II, resulting in decreased pre and after-load from venous and arteriodilation. It also decreases salt and water retention by reducing aldosterone production. It is often used in combination with loop diuretics in the treatment of congestive heart failure. ACE inhibitors may exacerbate pre-renal azotemia in hypotensive or renal compromised rabbits. ACE-inhibitors are beneficial in cases of hypertension and in some cases of chronic renal failure. Benazepril (another ACE-inhibitor) exerts its effects in a similar manner to enalapril. It is also commonly used in association with loop diuretics in the treatment of congestive heart failure. Benazepril is also useful in the treatment of hypertension and chronic renal insufficiency. Benazepril has significant hepatic metabolism and it is metabolized to the active form benazeprilat as in dogs.

- Taurine, given orally @(100 mg daily , may have an added advantage of slowing down the rapid progression of heart failure (Takihara, 1986).
- Pimobendan has inotropic, vasodilatory and anti-inflammatory effects and is recommended in the management of systolic dysfunction and myocardial failure in rabbits also (Periaut, 2009). Its dose rate in rabbits is 0.1–0.3 mg/kg PO once or twice daily. The drug should be used empty stomach as its bioavailability is reduced in the presence of food.

Drugs Used in the management of Heart Failure in Rabbits

Lidocaine - It is used in the management of ventricular arrhythmias @ 0.5-1.0 mg/kg IV

Atropine - It is recommended for bradycardias in canines. It is ineffective in many rabbits owing to the presence of atropinesterases (Varga,2014).

Glycopyrrolate - It is recommended in the management of bradycardias @ 0.01-0.1 mg/kg IV/SC/IM.

Atenolol - It is a beta blocker. It is used to manage tachycardia @ 0.5-2.0 mg/kg PO OD.

Furosemide –It is a loop diuretic and is used to reduce pulmonary edema @ 0.3- 4.0 mg/kg IV/IM/SC /PO OD/BID

Diltiazem - It is a calcium channel blocker and is recommended in the management of atrial arrhythmias. Its dose rate is 0.5-1.0 mg/kg PO BID/OD

Enalapril - It is ACE inhibitor. Its dose rate is 0.1-0.5 mg/kg PO every 24 or 48 hrs.

Benazapril - ACE- inhibitor. 0.05 mg /kg PO OD (max. upto 0.1 mg/kg OD)

Glycerine nitrate or Nitro-glycerine 2% ointment - It is a vaso-dilator and used as an ointment @ 1/8" applied transdermally every 6 to 12 hours.

References

Houston,S.,Ming-Show Lee,P., Quesenberry,K. and Pilny,A. (2012). Cardiovascular Disease, Lymphoproliferative Disorders and Thymomas. In: Ferrets, Rabbits and Rodents: Clinical Medicine and Surgery. Carpenter,J. and Quesenberry,K. 3rd edn. St Louis: Elsevier Saunders, pp.257-268.

Pariaut, R. (2009). Cardiovascular physiology and diseases of the rabbit. Vet. Clin. North. Am. Exot. Anim. Pract. 2009;12:135–144. [PubMed] [Google Scholar]

Takihara, K., Azuma, J., Awata, N., Ohta, H., Hamaguchi, T., Sawamura, A., Tanaka,Y., Kishimoto, S., and Sperelakis. N. (1986). Beneficial effect of taurine in rabbits with chronic congestive heart failure. Am. Heart J. 112: 1278-84. PMID: 3788775 DOI: 10.1016/0002-8703(86)90360-1

Varga, M. (2014). Cardiorespiratory disease.In: Text book of Rabbit Medicine 2nd edn. Butterworth-Heinemann; Edinburgh: . [Google Scholar]

Vennen, K.M. and Mitchell,M.A. (2009). Rabbits: Common Disease Presentations. In : Manual of Exotic Pet Practice. Mitchell,M. and Tully,T. St Louis: Elsevier Saunders, pp.385-396.

Further Readings

Carpenter, J.W. (2005). Exotic Animal Formulary . 3rd edn. St. Louis, Mo. : Elsevier Saunders,

Fontes-Sousa, A.P., Moura, C., Carneiro, C.S., Teixeira-Pinto, A., Areias, J.C. and Leite-Moreira, A.F. (2009) Echocardiographic evaluation including tissue Doppler imaging in New Zealand white rabbits sedated with midazolam and ketamine. Vet.J. 181 326-331

14

Heart Failure in Pet Avian

Heart failure is the end stage of cardiac diseases. It occurs in pet birds when cardiac output is insufficient and unable to maintain arterial blood pressure. In birds right side heart failure is more common as compared to left heart failure.

Etiology

- Congenital heart diseases, degenerative heart diseases, ingestion of toxic materials, tumors and idiopathic causes may lead to heart failure in birds.
- Heart failure is also observed in pet birds on feeding of Avacado seeds and flesh, and chocolates.
- Atherosclerosis in old pssitacines may be the cause of sudden death. It is a chronic disease of arterial walls characterized by narrowing of the arterial lumen due to accumulations of cholesterol, fat, calcium and inflammatory cells.

Clinical Manifestations

- Symptoms of heart diseases are subtle in birds as they are able to mask symptoms until there is congestive heart failure (Welle, 2016).
- Symptoms of congestive heart failure are manifested as per the side of the heart involved. Right side heart failure is quite common in birds as compared to left heart failure.
- The most common clinical manifestations of right heart failure in birds are lethargy, exercise intolerance, hyporexia, distension of coelomic cavity, pericardial effusion, hepatic congestion, jugular distension and dyspnea owing to compression of air sacs (Beaufrère and Guzman, 2016). Coughing is usually not common as diaphragm is lacking in birds. Bluish tint to the skin around eyes especially in African grey parrots may be seen in case of heart failure.

- Intermittent paresis secondary to hypoperfusion of hind limbs has been reported in Amazon parrot with atherosclerosis ((Beaufrere *et al.*, 2001).

Diagnosis

- Ante mortem diagnosis of heart diseases in avian species is a challenging job for veterinarians because of some anatomical and physiological features unique to birds impending assessment of cardiovascular system. Heart size in birds can vary according to activity levels. Large birds have proportionately smaller heart in relation to body size than smaller birds (Smith *et al.,* 2000).
- Clinical examination, auscultation, radiography, electrocardiography and echocardiography are employed in evaluating birds with suspected heart failure. There is no palpable pulse and auscultation is difficult to evaluate.
- Echocardiography is very useful in detecting effusions. In psittacines and raptors, ventromedial window is generally used to obtain two chamber (left atrium and left ventricle as well as atrioventricular valve and small part of the right ventricle) and four chamber view (left atrium, left ventricle, right atrium, right ventricle, atrioventricular valve, and aortic root with aortic valve) of the heart.
- Radiography is commonly employed diagnostic modality in avian species .Radiographs are generally taken after sedation and are very useful in assessing cardiac silhouette, great vessels, hepatic congestion or ascites. Increased radio densities in the lungs, enlargement of liver silhouette, or ascites is common in avian with congested heart failure .Enlargement of heart shadow may be seen in cases with dilatation, hypertrophy, pericardial effusion or tumors. Width of cardiac silhouette is measured on ventro-dorsal radiographs.
- Arrhythmias and changes in chamber size can be detected by electrocardiographic examination. It is better if electrocardiographic examination is conducted under sedation.
- Blood pressure monitoring can be done by placing blood pressure cuff (size 30-40% of limb circumference) over superficial ulnar artery or cranial tibial artery using Doppler ultrasonic flow detector and sphygmomanometer (Fitzgerald and Beaufrere, 2016). Birds have much higher systolic pressure as compared to mammals. Blood pressure monitoring has limited use in pet birds.

Therapeutic Protocol

- Generally the principles of cardiac therapy in avian species are almost similar to mammals. Nevertheless few points are to be kept in mind while dealing with avian species. There are high grade alterations in cardiac compromised avian patients such as marked weakness, ascites, and emaciation and high grade circulatory disturbances. Effects and adverse effects of the drugs depend on the species of the bird.
- The information on the treatment of cardiac failure and cardiac diseases in pet avian is far less as compared to canines and felines.
- The birds should the treated in calm and temporarily darkened place.
- The birds with acute congestive heart failure require oxygen therapy to stabilize.
- Pericardial or peritoneal effusions need to be drained to reduce dyspnea. When pericardial effusions is of low grade, pericardiocentesis has a positive beneficial effect. Removal of the fluid should not be > 10 ml/kg b.wt./day otherwise hypovolemic shock may occur.
- Fluid therapy is essential to prevent circulatory collapse, shock and dehydration when diuretics or ACEI or ACE- blockers are used. Fluid may be given IV when bird is in shock, followed by subcutaneous or intraosseous route.
- Diuretics helps in decreasing smaller coelomic effusions and to reduce fluid overload. Furosemide is commonly used diuretic (@ 1.0 to 2.0 mg/kg PO OD/BID). Lories are sensitive to diuretics, so use diuretics in Lories with care. Anorexia, vomiting, hypovolemia and hypokalaemia are the common side effects of diuretic therapy with furosemide. Spironolactone (@1.0-2.0 mg/kg PO OD/BID) , a potassium sparing diuretic, has also been used successfully in psittacine birds.
- ACE-inhibitors are helpful in improving ventricular filling and cardiac output (Beaufrère and Guzman, 2016).Studies on the use of ACE-I are limited in birds. Renin-angiotensin system is little different in birds. When birds are treated with ACE-I, electrolyte level should be monitored. ACE-I do not cause substantial decrease in arterial blood pressure in birds.
- Pimobendan, a PDE-III inhibitor, has been used in the management of congestive heart failure for its +ve inotropic, calcium sensitizer, vasodilator, venodilator, arteriodilator properties without increasing the demand of oxygen. Pharmokinetic of single dose of pimobendan

has been studied in Amazon parrots (Guzman *et al.*, 2014) . But further studies evaluating the effect of oral administration of pimobendan in bird's species are needed for definite recommendations.

- Calcium channel blockers have not been fully evaluated in avian medicine.
- Digoxin (initial @0.02 mg/kg PO OD, maintenance @ 0.01 mg/kg PO OD) is indicated in the management of chronic heart failure accompanied by atrial fibrillations. Digoxin improves cardiac output, stroke volume, and arterial pressure towards normal and decreases venous pressure (Hamlin and Stalnakar, 1987). If there are ventricular arrhythmias, treatment with digoxin should not be initiated. Sign of toxicity are seen in Indian Hill Mynah when treated with usual dose rate of digoxin. Cardiac arrhythmias and gastrointestinal signs (anorexia) are some of the signs of toxicity of digoxin in birds. Digoxin therapy requires regular monitoring of serum digoxin level. Long term use of digoxin may become problematic as it is difficult to control effect and plasma levels and sudden death may be observed possibly due to development of arrhythmias.
- Supportive therapy with vitamins, electrolytes, amino acids and buffers may also be given as per merit of the case.
- In birds with arteriosclerosis, treatment with vitamin A,C, E and selenium is also recommended.
- Heart failure affects the functions of liver, kidney and lungs. Therefore an eye should be kept on the functioning of these organs during the treatment of heart failure. ACE-I affects kidney function.
- Birds with heart failure should be kept in a comfortable environment (preferable temperature around 25-26^{0}C and humidity around 60-65%).

Shock

- Shock is very common in pet avian.
- Physical or emotion trauma, stress of handling, terrorization by cat, blood loss, dehydration, severe acute infections or poisoning may lead to shock in pet birds.
- Shock may be mild or severe leading to collapse, coma or death.
- Handling a pet bird affected with shock may aggravate shock and lead to death. Hence, a bird with shock needs to be handled cautiously with

utmost care and owner should be well informed about the impending catastrophic result.

- Hypothermia (<40-42 ^{0}C), cool extremities, tachycardia, pale mucus membrane, weak pulse, slow capillary refill time, dullness, weakness, depression, sleepiness, fluffed up feathers, tachypnea, tail bobbing, open mouth breathing ,seizures, head turned towards wing, half closed/ closed eyes and falling on the floor may be observed in pet avian with shock (Varshney, 2021). Such birds should not be disturbed unnecessarily and handled minimally.
- Hypovolemic shock is commonly observed in birds. This type of shock is due to an inadequate blood volume. Absolute hypovolemia occurs due to hemorrhage from trauma, coagulopathies or gastrointestinal tract bleeding. While relative hypovolemia occurs due to severe dehydration from the GIT or with extreme loss of fluids into a third body space (e.g., abdominal ascites.). Birds are able to tolerate hemorrhage better as compared to mammals. PCV can provide an index of dehydration and dehydrated birds should be given intravenous or intraosseous fluids. Birds with hemorrhage should be given vitamin K, calcium and antibiotics as per necessity. Steroids administration in hemorrhage and hypovolemic shock is not currently recommended. Birds with packed cell volume < 20 % require blood transfusion. Homologous (same species) transfusion is always preferable as the red blood cells last longer (9–11 days). Anticoagulant citrate phosphate or a heparinized syringe and a blood filter must be used. A donor bird can safely give blood @ 1% of body weight, e.g., 5 ml from healthy 500 g bird (Jenkins, 1997).
- Skin tenting is not reliable test for assessing the status of hydration in the avian patient. A better option is to press basilica wing vein. In healthy subjects the vein is refilled immediately. If refilling takes > 2 seconds, it is suggestive of > 7% dehydration. A quick rule of thumb is that it can be assumed that a very sick bird is about 10% dehydrated
- If shock is due to hemorrhages, bleeding control should be given first priority.
- Birds with shock need be kept warm (25-30 ^{0} C) , quiet, secluded in humid and dimly lit environment.
- Gasping birds should be subjected to immediate oxygen therapy.
- Dehydrated birds need intravenous or intraosseous fluid therapy with warm lactated ringers @10 ml/kg. Basilic, right jugular (it is larger) or

metatarsal vein can be used for intravenous fluid transfusion. In small or collapsed birds an intraosseous catheter can be placed aseptically in either the distal ulna or the proximal tibiotarsus. Short spinal needles (20–22 g) can be used in large birds and injection needles of 22–25 g can be used in small birds. The needles can be placed under a light general anesthesia or under lidocaine (0.2 mg/ 100g diluted with saline) on the skin and periosteum. The fluid deficit can be calculated by the formula: Estimated dehydration % x body weight (g) = fluid deficit. Half fluid is given in the first 24hrs. The daily maintenance is then 75–100 ml/kg (Lichtenberger, 2007).

- Favorite familiar food should be provided to encourage feeding.
- Anorectic birds need fluid through gavage tube when they are stabilized.
- Antibiotic or antifungal drugs may be started as per suspicion, If there is history of animal attack injury, piperacillin or doxycycline may be given.
- Analgesia is vital in cases of trauma. Buprenorphine, butorphanol, carprofen or meloxicam may be used if there is no renal failure and dehydration.
- Avoid giving food and water to the injured birds in shock.
- Cardiac arrest has poor prognosis as direct cardiac compression is restricted because of the sternum. Cardiopulmonary resuscitation can be undertaken in birds with respiratory arrest in case of overdose of anesthetic by turning off the anesthesia, providing intubation if necessary and intermittent positive pressure ventilation (IPPV) every 5 seconds with 100% oxygen. When heart beat is inaudible sternum is compressed, ventilation is provided, atropine (@ 0.2 mg.kg intravenously) and epinephrine (0.01mg.kg diluted in saline via a catheter down the trachea) are given. A sterile endotracheal tube needs to be inserted and stitched into left abdominal air sac in case of tracheal obstruction.

Drugs Used in the Management of Heart failure in Pet Birds

Butorphanol @1.0 mg/kg IM

Buprenorphine @ 0.01-0.05 mg/kg IM

Carprofen @ 2.0-4.0 mg/kg IM/SC/PO BID

Digoxin @ 0.02-0.05 mg/kg PO BID Initial. Followed by 0.01-0.02 mg/kg PO BID

Enalapril @ 0.25 to 0.5 mg/kg /day PO

@ 1.25 mg/kg body mass PO BID in Pigeons and Amazon Birds

Furosemide @ 0.15–2 mg/kg, PO/SC/ IM, OD/BID

Pimobendan @ 0.25 mg/kg, PO BID

Ringers Lactate @ 10 ml/kg IV/IO

References

Beaufrere, H., Holder, K.A., Bauer, R., Schnellbacher, R., Pariaut, R., Tully Jr.,T.N., Wakamatsu, N. (2001). Intermittent Claudication-like syndrome secondary to atherosclerosis in a yellow-naped amazon parrot (Amazona ochrocephala auropalliata). J. Avian Med.Surg. 25:266-276.

Beaufrere, H. and Guzman, D.S.M. (2016). Disorders of the cardiovascular system. In: Avian Medicine. Samour, J. (ed.). 3rd edn. Missouri:Elsevier. pp 395-407.

Fitzgerald, B.C. and Beaufrere, H. (2016). Cardiology. In:Current Therapy in Avian medicine and Surgery. Speer, B.L. (ed.). Missouri:Elsevier. pp 252-328

Guzman, D.S., Beaufrere, H., KuKanich, B., Barker, S.A., Brandao, J., Paul-Murphy, J. and Tully Jr., T.N. (2014). Pharmacokinetics of a single oral dose of pimobendan in Hispaniolan

Amazon Parrots (Amazona ventralis). Avian Med.Surg. 28: 95-101. Doi:10.1647/2012-061. Hamlin, R.L. and Stalnaker, P.S. (1987). Basis for use of digoxin in small birds. J. Vet. Pharmacol. Therap.10:354-350.

Jenkins, J.R. (1997). Avian critical care and emergency medicine. Avian Medicine and Surgery. WB Saunders, Philadelphia .pp 839–863.

Lichtenberger, M. (2007). Shock and cardiopulmonary-cerebral resuscitation in small mammals and birds. Vet. Clin. Exot. Anim. 10:275–291.

Smith, F.M., West, N.H. and Jones, D.R. (2000). The cardiovascular system. In: Sturkie's Avian Physiology. Whittow, G.C. (ed.). 5th edn . San Diego: Academic Press. pp 41-231.

Varshney, J.P. (2021). Handbook of Exotic Pet Medicine. Brillion Publishing, New Delhi-110005 pp.114-115

Welle, K, R. (2016). Cardiac Disease. In: Blackwell's Five-minute Veterinary Consult: Avian. Graham, J.E.(ed). John Wiley and Sons .pp 52-53.

15

Heart Failure in Ferret

The ferret is a domestic species of small mustelid. Congestive heart failure is most common form of heart disease seen in aged ferrets. Heart failure in ferrets is due to heart diseases and clinical presentation varies from no clinical signs in initial stage to many signs of fulminant heart failure. Initially ferrets, being a prey mammals, show subtle signs. Signs of heart failure are expressed in ferrets only in the advanced stage of the heart disease. Dilated cardiomyopathy is common in middle aged to older ferrets but hypertrophic cardiomyopathy and acquired valvular disease are also seen in ferrets .Ferrets are susceptible to heart worm disease. Congested heart failure is the end stage manifestation of these cardiac diseases.

Etiology

- Aortic and mitral valve defects are common cause of heart failure in ferrets (Orcutt and Malakoff, 2006).
- Dilated and hypertrophic cardiomyopathies have also been described as a cause of heart failure in ferrets.
- Heartworm disease (due to *Dirofilaria immitis*) in ferret is the major cause of heart failure
- Heart failure may also be seen in ferret secondary to kidney disease.

Clinical Manifestations

Signs of heart failure in ferrets are not expressed until late into the progression of cardiac disease. The following signs have been described in ferrets with heart failure and seem related to forward and backward failure. Congestive heart failure is an insidious and progressive condition.

- Anorexia
- Marked weakness
- Exercise intolerance

- Progressive loss in body weight
- Pale mucus membranes
- Increased capillary refill time
- Dyspnea/ tachypnea more predominant in cases with thoracic effusions.
- Gasping/ wheezing during or after strenuous activity.
- Prominent hind limb weakness
- Ascites (potbelly), limb edema or retention of fluid in chest cavity.
- Pain due to fluid retention
- Coughing is uncommon
- Cardiac murmur are associated with valvular insufficiency and dilated cardiomyopathy.

  Murmurs are uncommon with hypertrophic cardiomyopathy.
- Cardiac arrhythmias

Diagnosis

- Clinical symptoms can arise suspicion of heart failure.
- Ferrets have a more globoid cardiac silhouette compared to dogs and cats. Contralateral radiographs of thorax under sedation may show enlarged cardiac silhouette touching the sternum and elevated trachea consistent with cardiac diseases (Morrisey and Kraus, 2012). However, heart size can be evaluated with a modified vertebral heart score as described by Stepien *et al*. (1999).
- Echocardiography can be helpful in detecting pericardial effusions, and valvular defects.
- Electrocardiography can be employed to reveal arrhythmias and conduction disturbances.
- Enlarged globoid heart with increased sternal and diaphragmatic contact is evident in ferrets with dilated cardiomyopathy. Fluid in lungs is the common radiographic finding in ferrets with congestive heart failure. Heart is not visualized in radiographs when pleural effusions are present. Care should be exercised in interpreting severe pleural effusions as there may be an imperceptible mass. Various degree of pulmonary edema may also be visible in ferrets with congestive heart failure.

- For the diagnosis of heartworm disease, ELISA based antigen snap test or Polymerase Chain Reaction test can be performed.

Treatment

Principles of treatment of heart failure (oxygenation, reducing the preload and after load) in ferrets are similar to canines.

- Furosemide (2.0-3.0 mg/kg IM/IV BID/TID), a loop diuretic, is of great value in reducing preload in fulminant heart failure. In long term use dose of furosemide should be reduced to 1.0-2.0 mg/kg PO BID.
- Angiotensin converting enzyme inhibitors (ACE-I) are used to reduce the afterload by peripheral vasodilation. Enalapril is commonly used for this purpose @ 0.5 mg/kg PO at 48 hrs interval. Ferrets are sensitive to hypotension, therefore requiring constant monitoring for the signs of lethargy.
- The use of pimobendan as a treatment for DCM in ferrets has been mentioned anecdotically (Delprat, 2005). The drug is well tolerated in ferrets (Hermans *et al.*, 2008). However, more studies are warranted on the use of pimobendan in the management of heart failure in ferrets.
- Beta-blockers, calcium channel blockers and digoxin can be used to manage arrhythmias. Digoxin can be used in cases of dilated cardiomyopathy @ 0.01 mg/kg PO OD as initial dose. The dose can be gradually increased to twice daily.
- Stress free environment and complete rest is necessary for better effect of treatments.

Prognosis

- Long term prognosis in ferrets with heart failure is guarded to poor.
- Prognosis in cases of congestive heart failure associated with heart worm disease is grave. But this is only condition that can be prevented (Antinoff *et al.*, 2000).

References

Antinoff, N., Garner , M., and Burgess, M. (2000). A Review of Diseases of the Ferret. Exotic DVM 2 (2): 33 – 40.

Delprat C. (2005). Un cas de cardiomyopathie chez un furet: extrapolation des thérapeutiques conventionelles de CMD canines. In: Proceedings congrès AFVAC, December 2005, Toulouse

Hermans, K., Geerts, T., Cauwerts, K., Salheen, M.T.S. and Baert, K. (2008). Tolerability of pimobendan in the ferret (Mustelaputorius furo). Vlaams Diergeneeskundig Tijdschrift 78 : 53-55.

Morrisey, J. and Kraus, M. (2012). Cardiovascular and Other Diseases. In: Ferrets, Rabbits and Rodents: Clinical Medicine and Surgery. Carpenter,J. and Quesenberry,K. 3rd edn. St Louis: Elsevier Saunders, pp.62-77.

Orcutt,C. and Malakoff, R. (2006). Ferrets: Cardiovascular and Respiratory System Disorders. In: BSAVA Manual of Roddents and ferrets, Gloucester. Meredith, A. and Lord, B. British Small Anim. Vet. Assoc., pp205-213.

Stepien, R.L., Benson, K.G. and Forrest, L.G. (1999). Radiographic measurement of cardiac size in normal ferrets. Vet.Radiol.Ultrasound, 40: 606-610.

16

Heart Failure in Man

Heart failure is a common progressive and complex clinical syndrome with high morbidity and mortality in humans. It is mostly due to cardiac involvement. Clinical signs and symptoms developed in heart failure cases are the manifestations of body's compensatory efforts to meet chronic inadequate cardiac output as seen in other species of animals also.

The information on heart failure in man in this chapter is not claimed to be complete or exhaustive. The purpose of this chapter is very restrictive to provide just some information for comparison purpose.

Causes of Heart Failure

- Heart diseases are main etiological factor for heart failure in humans.
- Coexistence of myocardial disease and systemic hypertension is very common in human cases of heart failure.
- Rheumatic valvular /or and myocardial disease may also lead to heart failure.
- Ischemic heart diseases (IHD) and primary myocardial inflammatory heart diseases also lead to heart failure in man.
- Coronary artery disease (infarction) is the most common cause of acute heart failure.
- Hypertension may lead to heart failure by increasing resistance to systolic ejection.
- Heart failure may also occur when stroke volume is increased in aortic incompetence and /or mitral incompetence.
- Dysrhythmia particularly prolonged tachycardia can also lead to heart failure in infants.
- Persons with diabetes mellitus, coronary artery disease, hypertension, on cardio toxic drugs, alcohol abuse, or with a family history of cardiomyopathy have a high risk of heart failure.

Types of Heart Failure

- Heart failure has been categorized in two main groups as (i) heart failure with reduced left ventricular ejection fraction (HE-rEF) and (ii) heart failure with preserved left ventricular ejection fraction (HF-pEF) .
- Heart failure with reduced left ventricular ejection fraction (HF-rEF)- This type of heart failure is a systolic failure in which left ventricle is enlarged and its contractility is reduced. The main echocardiographic feature of HF-rEF is reduced ejection fraction of left ventricle (<40%). Systolic dysfunction (reduced myocardial contractility) is commonly associated with myocardial infarction, idiopathic dilated cardiomyopathy, hypertension and valvular disease. Systolic dysfunction can lead to high cardiac output heart failure (as seen in cases of severe anemia, arteriovenous malformations with shunting of blood, or hyperthyroidism), low cardiac output heart failure, right heart failure, left heart failure or biventricular heart failure.
- Heart Failure with preserved left ventricular ejection fraction (HF-pEF) - This type of heart failure is a diastolic failure in which heart contracts normally but fails to relax owing to thickened and stiff ventricular wall. Left ventricular ejection fraction is >50% in this type of heart failure. Diastolic dysfunction is commonly associated with coronary artery disease, hypertension, aging and infiltrative cardiomyopathy.
- Patient having left ventricular ejection fraction in between 40 to 49% are grouped as heart failure with mid-range ejection fraction. Prevalence rate, pathophysiology, treatment and prognosis of heart failure cases with mid- range ejection fraction is still not clear (Ponikowski *et al.*, 2016).

Pathophysiology of Heart Failure

- Major pathophysiology of heart failure in humans is almost similar to that described in the dogs.
- Congestive heart failure, a syndrome, is characterized by pressure and volume overload, reduced myocardial contractility, reduced cardiac output, or high output failure.
- In the usual form of heart failure, the heart muscle has reduced contractility. There is alteration in primary determinants of left ventricular (LV) performance.

- Pulmonary congestion and dyspnea are due to increased preload or left atrial filling pressure .
- Although systemic blood pressure is often reduced, an increase in systemic vascular resistance (afterload) causes further reduction in cardiac output.
- Compensatory mechanism associated with an increase in sympathetic tone and circulating catecholamine in heart failure increases heart rate.
- Coronary disease in patients increases myocardial oxygen demand.
- Enlargement of heart size in long standing cases of heart failure is deleterious because cardiomegaly increases wall tension and also myocardial oxygen consumption.
- Cardiac muscle hypertrophy is a compensatory attempt of the body to maintain cardiac function.
- A rise in LV filling pressure is also a compensatory effort to optimize cardiac performance.
- An increase in peripheral arterial-venous oxygen extraction is again a compensatory effort to maximize the oxygen delivered for a given cardiac output.
- Although these compensatory mechanisms are initially helpful, excessive increase of many of them (catecholamine and systemic vascular resistance) may actually be detrimental.
- Increased sympathetic stimulation also causes more secretion of vasopressin from the posterior pituitary. Increased vasopressin in turn results into fluid retention at the kidneys, increase in the blood volume and blood pressure.
- Inability of kidneys to dispose of sodium and water, further increases edema.
- Hypo perfusion of kidneys stimulates the release of renin which catalyzes the production of the angiotensin.
- Angiotensin and its metabolites cause further vasoconstriction, and stimulate increased secretion of the steroid aldosterone from the adrenal glands promoting salt and fluid retention at the kidneys.
- Persisting high levels of circulating neuroendocrine hormones (catecholamine, renin, angiotensin, and aldosterone) causes structural remodeling of the heart over long time.

- Reduced perfusion of skeletal muscle causes atrophy of the muscle fibers leading to weakness and increased fatigue.
- In right-sided heart failure, edema starts appearing in the ankles. Later on ascites may also develop.
- In left-sided heart failure edema occurs in the lungs (cardiogenic pulmonary edema). It is the cause of dyspnea, orthopnea and paroxysmal nocturnal dyspnea.
- Cold extremities, cyanosis, weakness, dizziness and fainting is due to decreased cardiac output and poor peripheral perfusion.

Symptoms

Clinical signs in cases of heart failure depend on the side of the heart affected (right heart failure or left heart failure or both).Most of the cases of left heart failure are due to ventricular failure. Common symptoms of heart failure in man are given below.

- Shortness of breath is marked with activity or on lying down.
- There is a feeling of fatigue and weakness.
- Swelling can be conspicuous in legs, ankles and feet.
- There may be variation in heart rate. Heart rate may be fast or irregular.
- Ability to do exercise and or to perform works is reduced.
- Persistent cough or wheezing with white or pink blood-tinged mucus is disturbing sign.
- Ascites may be visible.
- Fluid buildup may increase weight.
- Nausea and lack of appetite becomes a common feature in standing cases of heart failure.
- There may be difficulty in concentrating thoughts.
- In many cases of heart failure alertness may also be decreased.
- Chest pain is a prominent symptom in cases of heart failure caused by a heart attack due to ischemic heart disease / infarction.

Symptoms Associated with Left Ventricular Failure (LVF)

- Dyspnea and nocturnal dyspnea is common clinical signs in LVF

- Coughing ,sometimes associated with hemoptysis, is also seen in humans with left heart failure
- In late stage of LVF, breathlessness increases with speech, emotion and with slightest effort.
- In advance cases breathing becomes quick, troublesome and shallow.
- Patients with LVF experience more dyspnea in recumbancy (when lying in bed) and become comfortable by sitting or standing. Paroxysmal cardiac dyspnea occurs at night particularly in horizontal posture.
- Anxiety of nocturnal attacks of dyspnea further deteriorate the condition.
- Pulmonary edema becomes more conspicuous in advance cases of LVF
- Precordial pain is also observed in many patients with LVF.
- Noisy respiration accompanied by coughing and expectoration of much watery mucus often pink colored can be seen in many cases.
- Cyanosis develops due to asphyxia caused by transudate.
- Cheyne - Stokes respiration can be seen in late stage of left ventricular failure especially in elderly patients.
- Tachycardia is common in patients with LVF.
- Cardiac enlargement is due to left ventricular hypertrophy in chronic heart failure cases
- During attack of left ventricular failure (cardiac asthma, pulmonary edema), the blood pressure is often elevated.
- Pallor skin and coldness of extremities is due to peripheral constriction in cases of left heart failure.
- Great left ventricular enlargement may be associated with the picture of right ventricular failure and the patient is able to lie flat.

Symptoms Associated with Right Ventricular failure

- Pulmonary hypertension (PH) is the common cause of right ventricular failure.
- Congestion due to left heart failure is another cause of right ventricular failure.
- Severity of dyspnea begins to reduce as right ventricular failure begins.
- Fatigue is more prominent.

- Hepatic congestion may cause discomfort in the right hypochondrium.
- Anorexia and vomiting are also seen.
- Right heart failure may lead to cardiac cirrhosis as seen in some patients.
- Edema may be seen in dependent parts, at the ankles, over the sacrum, at the back of thigh and in genitalia.
- Pleural effusions, ascites and pericardial effusion are also commonly observed in cases of right ventricular failure.
- Peripheral arteriolar constriction and stasis of blood may cause peripheral cyanosis.

Stages of Progression of Heart Failure

As per guidelines from the American College of cardiology and American Heart Association, progression of heart failure has been identified in four stages (Hunt *et al.*,2001) as follows:

Stage A. Patients have no structural abnormality but are at high risk.

Stage B. Patients are asymptomatic but have structural heart disease.

Stage C. Patients have structural abnormalities with past or present heart failure.

Stage D. Patients have end stage heart failure and require mechanical circulatory support, infusion of inotropic drugs, cardiac transplantation.

Diagnostic Approach in Heart Failure

- Diagnosis of heart failure depends on focused evaluation of history and physical examination, chest radiograph, , an electrocardiogram and an echocardiogram.
- Echocardiogram plays a confirmatory role in the diagnosis of heart failure. It is the most suitable test that provides immediate information about systolic and diastolic functions of both the ventricles and atrium , chamber size, wall thickness, and valvular insufficiencies. Transthoracic echocardiography, strain measured by echocardiography, cardiovascular magnetic resonance (CMR), computed tomography (CT scan) are the other emerging cardiac imaging techniques to visualize and quantify cardiac structure and functions in cases of heart failure.
- Radiographic examination is helpful in revealing vascular congestion in lungs. During an early stage of left ventricular failure, slight increase in

the upper lobe veins may be visualized. In advanced cases , extensive congestion in lung field may be detected. In pulmonary edema, massive shadows extending from hilar regions over the mid zones of the lung leaving apices and base are quite common in radiographs. Heart shadow is usually enlarged in well-established cases of heart failure.

- Urine examination may reveal oliguria , concentrated urine with low sodium chloride and slight albuminuria.
- With low urine output, blood urea is moderately increased.
- In the absence of dyspnea, and an abnormal chest radiograph or an electrocardiogram, heart failure is highly unlikely.
- Radionuclide angiography or contrast cineangiography are the other imaging techniques being used in suspected cases of heart failure with no significant echocardiographic changes..
- Patients with confirmed heart failure are subjected to additional tests viz. complete blood count, blood glucose, liver function test, kidney function test, lipid profile, serum electrolytes, thyroid stimulating hormone level and urinalysis for assessing the associated effects of heart failure.
- Estimation of B-type natriuretic peptide (BNP) and N-terminal pro-BNP (NT-pro BNP) have diagnostic potential for detecting both HFrEF and HFpEF in humans also. Heart failure is considered highly unlikely if NT- pro BNP is below the cut off value. Roberts *et al.* (2015) found that the use of NPs was more suitable for excluding HF owing to low positive predictive value of NP (0.66 to 0.67).
- Soluble suppression of tumorigenicity-2(ST-2), galectin-3, and high sensitivity cardiac troponin are the other novel biomarkers being investigated for detecting myocardial injury and fibrosis and for additive risk stratification in heart failure patients (de Boer *et al.,* 2015, Yancy *et al*, 2018).

Developments in Therapy of Heart Failure in Man

- Before 1980, the treatment of heart failure cases was mainly based on bed rest, fluid restriction, digitalis and diuretics.
- During 1970 and 1980, inotropes and vasodilators were added in the treatment of heart failure.
- Since 1980, heart failure has been understood as a neurohormonal disease. This led to improve cardiac function by RAAS inhibitors and by sympathetic activation.

- Administration of ACE-inhibitors and beta- blockers in the management of heart failure has reduced mortality and improve LV function in humans.
- New drugs- Mavacamten and omecamtiv mecarbil are the new drugs that acts on cardiac myosin are being investigated .Mavacamten acts as an inhibitor of cardiac myosin ATPase and reduces cardiac contractility has been investigated in humans with obstructive HCM (Varian and Tang, 2017). Omecamtiv mecabil (OM), a selective cardiac myosin activator has shown positive results in chronic heart failure (Teerlink *et al.*, 2016).

  SGLT-2 inhibitor such as Empagliflozin and canagliflozin have shown reduction in heart failure hospitalization rate and cardiovascular death in patients with type 2 diabetes mellitus Liraglutide, a glucagon like peptide 1 analog, has also shown decrease in all-cause mortality and cardiovascular deaths. It is an antidiabetic drug with additional cardiovascular benefits
- Physical and mental rest is necessary to reduce work load of the heart.
- Oxygen therapy is indicated when oxygen level is reduced because of pulmonary insufficiency

New Therapy

- Anti-inflammatory therapy- Release of various cytokines in heart diseases and heart failure has been documented in various studies. Tumor necrosis factor α (TNF-α), transforming growth factor β (TGF-β), and interleukins (IL-1, IL-4, IL-6, IL-8, and IL-18) have been involved in the development of various inflammatory cardiac pathologies.

  Anti-inflammatory therapy using canakinumab, a monoclonal antibody targeting IL-1β,has been reported to lower recurrent cardiovascular events in humans with previous myocardial infarction(MI) but it caused serious infectious complications (Ridker *et al.,* 2017). Anakinra, an IL-1 receptor antagonist, has shown some beneficial results in recently decompensated systolic HF patients (Van Tassell *et al.,* 2017a)], but no effect in HFpEF patients (Van Tassell, *et al*., 2017b).
- Gene therapy- Cardiac gene therapy, involving the production of proteins with curative efficacy by transferring specific exogenous genes, has been proposed as an important alternative therapeutic approach (Gabisonia and Recchia, 2018). The major targets of gene therapy are the β-adrenergic system, angiogenesis, cytoprotection, and stem

cell homing. Regardless of the etiology, it has been demonstrated that SERCA2a is deficient in experimental models of HF, and correction of SERCA2a deficiency can improve calcium influx and cardiac function. However, researches have shown that administration of SERCA2a does not improve the clinical course of HFrEF patients (Greenberg *et al.*, 2016).

- Stem cell therapy-Present therapeutic intervention in heart failure cases do not replace or regenerate damaged cardiomyocytes. This fact has led to the development of stem cell therapy. Use of stem cells has shown to improve myocardial function and ventricular remodeling, but lacking consistency in results (Tompkins *et al.*, 2018). Human pluripotent stem cells (hPSCs) have emerged to replace embryonic stem cells.

Routine Treatment Protocol

The following drugs are routinely used in the management of heart failure in humans.

- Diuretics- Furosemide is the drug of first choice.
- ACE- inhibitors-Enalapril, captopril and lisinopril are commonly used drugs.
- Angiotensin-II receptor Blocker-Losartan, valsartan, candesartan are the drugs in this group. These drugs may be used as an alternative to ACE-Inhibitors.
- Beta Blockers- Carvedilol, metoprolol, and bisoprolol are the drugs in this group used to reduce blood pressure and heart rate. These drugs have shown positive effect in cases of heart failure in humans.
- Aldosterone Antagonist- Spironolactone and eplerenone are the drugs which are aldosterone receptor blocker. These are potassium sparing and effective in people with severe systolic heart failure. They may increase potassium level. Therefore monitoring of potassium level should be done.
- Inotropes- These drugs are used to increase myocardial contractility and to slow down heartbeat. Digoxin has been the drug of choice in this group. It is also recommended if there are atrial fibrillations. Pimobendan is a novel drug being advocated in the management of congestive heart failure in veterinary medicine. It is being used in the management of congestive heart failure in humans in Japan (Kato, 1997). It is effective in patients with chronic heart failure but is less effective in patients

with old myocardial infarction (OMI) than in patients with dilated cardiomyopathy (DCM) or other heart diseases (Takeda *et al.*, 2001).

- Vasodilators-Hydralazine and isosobide di-nitraete are added to treatment protocol of severe heart failure not adequately responding to ACE-inhibitors and beta blockers.
- Surgical procedures viz. coronary bypass surgery, heart valve repair or replacement, implantable cardioverter-defibrillators(ICDs), cardiac resynchronization therapy (CRT),ventricular assist devices (VADs) and heart transplant are advanced options to treat the underlying problems in cases of heart failure in humans.

References

de Boer, R.A., Daniels, .LB., Maisel, A.S. and Januzzi, J.L. Jr.(2015). State of the art: newer biomarkers in heart failure. Eur. J. Heart Fail. 2015;17: 559–569. [PubMed] [Google Scholar] Gabisonia, K. and Recchia, F.A. (2018). Gene therapy for heart failure: new perspectives. Curr. Heart Fail. Rep. 15:340–349. [PMC free article] [PubMed] [Google Scholar]

Greenberg , B., Butler, J., Felker, G.M.,Ponikowaski,P.,Voors, A.A.,Desai,A.S., Barnard, D., Bouchard,A.,Jaski, B., Lyon,A.R.,Pogoda, J.M.,Rudy, J.J. and Zsebo,K.M. (2016). Calcium upregulation by percutaneous administration of gene therapy in patients with cardiac disease (CUPID 2): a randomised, multinational, double-blind, placebo-controlled, phase 2b trial. Lancet. 387:1178–1186. [PubMed] [Google Scholar]

Hunt, S.A., Baker, D.W., Chin, M.H., Cinquegrani, M.P., Feldman, A.M., Francis, G.S., Ganiats, T.G., Goldstein, S., Gregoratos, G., Jessup, M.L., Noble, R.J., Packer, M., Silver, M.A., Stevenson, L.W., Gibbons, R.J., Antman, E.M., Alpert, J.S.,

Faxon, D.P., Fuster, V., Jackobs, A.K., Hiratzka, L.F., Russell, R.O. and Smith Jr., S.C. (2001). ACC/AHA guidelines for the evaluation and management of chronic heart failure in the adult: executive summary. A report of the American College of Cardiology/American Heart Association Task Force on Practice Guidelines (Committee to revise the 1995 Guidelines for the Evaluation and Management of Heart Failure). J Am Coll Cardiol. 38:2101–13.

Kato,K. (1997). Clinical efficacy and safety of pimobendan in the treatment of heart failure-Experience in Japan. Cardiology 88;(supplement2)28-36.Doi:10.1159/0001777482.

Ponikowski, P., Voors, A.A., Anker, S.D., Bueno, H., Clelang, J.G.F., Coats, A.J.S., Falk, V., Gonjal Juanatey, J.R., Harjola, V., Jankowska, E.A., Jessup, M., Linde, C., Nihoyannopoulos, P., Parassis, J.T., Pieske, B., Riley, J.P., Rosano, G.M.C., Ruilope, L.M., Ruschitzka, E., Rutten, F.H., and Van der Meer, P. (2016). ESC guidelines for the diagnosis and treatment of acute and chronic heart failure: the task force for the diagnosis and treatment of acute and chronic heart failure of the European Society of Cardiology (ESC) developed with the special contribution of the Heart Failure Association (HFA) of the ESC. *Eur Heart J.* 37:2129– 2200. [PubMed] [Google Scholar]

Ridker, P.M., Everett, B.M., Thuren, T.MacFadyen, J.G., Chang, W.H., Ballantyne, C., Fonseca, F., Nocholau, J., Koenig, W., Anker, S.D., Kastelein.J.J.P. and Cornel.J.H. (2017). Antiinflammatory therapy with canakinumab for atherosclerotic disease. N. Engl. J .Med. 377:1119–1131. [PubMed] [Google Scholar

Roberts, E., Ludman, A.J., Dworzynski, K., AlMohammad, A., Cowir, M.R., McMurray, J.J.V. and Mant, J. (2015). The diagnostic accuracy of the natriuretic peptides in heart failure: systematic review and diagnostic meta-analysis in the acute care setting. *BMJ.* 2015;350:h910. [PMCfree article] [PubMed] [Google Scholar]

Takeda,N., Hayashi,Y., Arino,T., Takeda,A. and Noma,K. (2001). Effect of pimobendan in patients with chronic heart failure. Exp. Clin. Cardiol. 6 ;195-199.

Teerlink, J.R., Felke, G.M., McMurray, J.J., Solomon, S.D., Adams Jr., K.F., Cleland, J.G.F., Ezekowitz, J.A., Goudev,A., Macdonald, P., Metra, M., Mitrovic, V., Ponikowski, P., Serpyits, P., Spinar,J., Tomcsanyi, J., Vandekerckhove, H.J., Voors, A.A., Monsalvo, M.L. Johnston, J., Malik, F.I. and Honarpour, N. (2016). Chronic oral study of myosin activation to increase contractility in heart failure (COSMIC-HF): a phase 2, pharmacokinetic, randomised, placebo-controlled trial. Lancet 388:2895–2903. [PubMed] [Google Scholar]

Tompkins, B.A., Rieger, A.C., Florea, V., Banerjee,M.N., Natsumeda, M., Nigh, E.D , Landin, A.M., Rodriguez, G.M., Hatzistergos, K.E.,Schulman, I.H. and Hare, J.M. (2018). Comparison of mesenchymal stem cell efficacy in ischemic versus nonischemic dilated cardiomyopathy. J. Am. Heart Assoc. 7: e008460. [PMC free article] [PubMed] [Google Scholar] Van Tassell, B.W., Canada, J., Carbone, S., Trankle, C., Buckley, L., Erdle, C.O., Abouzaki, N.A., Dixon, D., Kadariya, D., Christopher, S., Schatz, A., Regan, J., Viscusi, M., Buono, M.D., Melchior, R., Mankad, P., Lu, J., Sculthorpe, R , Biondi-Zoccai, G., Lesnefsky, E., Arena, R. and Abbate, A. (2017a). Interleukin-1 blockade in recently decompensated systolic heart failure: results from REDHART (Recently Decompensated Heart Failure Anakinra Response Trial).Circ. Heart Fail. 10:e004373. [PMC free article] [PubMed] [Google Scholar]

Van Tassell, B.W., Buckley, L.F., Carbone, S., Trankle, C.R., Canada, J.M., Dixon, D.L., Abouzaki, N., Oddi-Erdle, C., Biondi-Zoccai, G., arena, R. and Abbate, A. (2017b). Interleukin-1 blockade in heart failure with preserved ejection fraction: rationale and design of the Diastolic Heart Failure Anakinra Response Trial 2 (D-HART2) Clin. Cardiol 40:626–632. [PMC free article] [PubMed] [Google Scholar]

Varian,K. and Tang,W.H.W.(2017). Therapeutic stratergies targeting inherited cardiomyopathies. Curr. Heart Fail. Rep. 14: 321-330.

Yancy, C.W., Januzzi, J.L.Jr., Allen, L.A., Butler, J., Davis, L.l., Fonarow, G.C., Ibrahim, N.E., Jessup, M., Lindenfeld, J., Madox, T.M., Masoudi, F.A., Motiwal , S.R., Patterson,J.H., Walsh , M.N. and Wasserman, A. (2018). ACC expert consensus decision pathway for optimization of heart failure treatment: answers to 10 pivotal issues about heart failure with reduced ejection fraction: a report of the American College of Cardiology task force on expert consensus decision pathways. J.Am.Coll.Cardiol. 71: 201–230. [PubMed] [Google Scholar]